AGRICULTURE ISSUES AND POLICIES

T0100218

LAND USE CHANGES

MANAGEMENT AND APPLICATIONS

AGRICULTURE ISSUES AND POLICIES

Additional books and e-books in this series can be found
on Nova's website under the Series tab.

AGRICULTURE ISSUES AND POLICIES

LAND USE CHANGES

MANAGEMENT AND APPLICATIONS

VINÍCIUS SANTOS ALVES
EDITOR

nova
science publishers
New York

Copyright © 2020 by Nova Science Publishers, Inc.

All rights reserved. No part of this book may be reproduced, stored in a retrieval system or transmitted in any form or by any means: electronic, electrostatic, magnetic, tape, mechanical photocopying, recording or otherwise without the written permission of the Publisher.

We have partnered with Copyright Clearance Center to make it easy for you to obtain permissions to reuse content from this publication. Simply navigate to this publication's page on Nova's website and locate the "Get Permission" button below the title description. This button is linked directly to the title's permission page on copyright.com. Alternatively, you can visit copyright.com and search by title, ISBN, or ISSN.

For further questions about using the service on copyright.com, please contact:
Copyright Clearance Center
Phone: +1-(978) 750-8400   Fax: +1-(978) 750-4470   E-mail: info@copyright.com.

NOTICE TO THE READER

The Publisher has taken reasonable care in the preparation of this book, but makes no expressed or implied warranty of any kind and assumes no responsibility for any errors or omissions. No liability is assumed for incidental or consequential damages in connection with or arising out of information contained in this book. The Publisher shall not be liable for any special, consequential, or exemplary damages resulting, in whole or in part, from the readers' use of, or reliance upon, this material. Any parts of this book based on government reports are so indicated and copyright is claimed for those parts to the extent applicable to compilations of such works.

Independent verification should be sought for any data, advice or recommendations contained in this book. In addition, no responsibility is assumed by the Publisher for any injury and/or damage to persons or property arising from any methods, products, instructions, ideas or otherwise contained in this publication.

This publication is designed to provide accurate and authoritative information with regard to the subject matter covered herein. It is sold with the clear understanding that the Publisher is not engaged in rendering legal or any other professional services. If legal or any other expert assistance is required, the services of a competent person should be sought. FROM A DECLARATION OF PARTICIPANTS JOINTLY ADOPTED BY A COMMITTEE OF THE AMERICAN BAR ASSOCIATION AND A COMMITTEE OF PUBLISHERS.

Additional color graphics may be available in the e-book version of this book.

Library of Congress Cataloging-in-Publication Data

ISBN: 978-1-53617-032-0

Published by Nova Science Publishers, Inc. † New York

CONTENTS

Preface   **vii**

Chapter 1   Applications of Land Use Change Prediction
                 Models (With Remote Sensing and Geographic
                 Information System Approach)   **1**
                 Jafar Jafarzadeh

Chapter 2   Land Concentration or Land Grabbing:
                 Multidimensional Issues Challenging
                 Ukrainian Agrarian Sector   **25**
                 Olena Borodina and Viktor Yarovyi

Chapter 3   Soil Functions and Land Management:
                 Potential Synergies and Tradeoffs at the
                 Tropical Agricultural Frontier (Brazil)   **59**
                 Maria Victoria R. Ballester, Andrea S. Garcia,
                 Rodnei Rizzo, Vivian M. F. Vilela,
                 Érica S. Nakai, Mayra de F. Preto,
                 Laura P. Casarin and Daiana M. Tourne

Contents

Chapter 4   Beginning of Desertification in the
                  Southern Buenos Aires and the Predictability
                  of Soil Water Content   **83**
                  Luciana Stoll Villarreal, Marcela Hebe González,
                  Alfredo Luis Rolla
                  and María Elizabeth Castañeda

Index   **109**

Related Nova Publications   **117**

PREFACE

Land use is one of the most important aspects of the study of natural resources management and environmental change. Today, recognizing these changes is possible by comparing the differences between images taken in several specified periods in a specific region. The details of these changes are Nunderstandable by mapping the components of land resources into more than one period.

Next, agricultural land use in Ukraine is explored. Providing general characteristics of processes in the Ukrainian agrarian sector witnessed over the past two decades, the authors estimate long-term trends and recent changes in agricultural land use and describe their drivers.

A framework is provided which addresses sustainable land management through ecosystem services provided by healthy soils. This conceptual tool is designed to support policy makers in the management of five selected soil functions and demands: productivity, water availability, nutrients, carbon sequestration and biodiversity.

In the closing study, the authors design a statistical model using atmospheric forcing to predict soil water storage for spring. The analysis of the efficiency of different models takes into account the adjusted squared correlation coefficient and cross-validation coefficient values.

Chapter 1 - Land use is one of the most important aspects of the study of natural resources management and environmental change review. Land use refers to human methods and purposes for resources. Identifying the timely and accurate changes in land use of the founder is a better way to understand the interactions of human and land resources. Recognizing these relationships leads to the management and sustainable use of these resources. Coverage and land use have two different meanings. Climate change, deforestation, flood and sediment increase, pollution, urban growth, desertification, soil erosion and... is all of the consequences of development without planning and regardless of the environmental impacts of land-use change. The pattern of coverage and land use of an area, the output of the relationship between nature and socioeconomic factors, their application by humans, in the dimension of time and place. Modified land is a land that has changed in terms of climate, topography, soil characteristics and land use concerning their previous condition. The process of these changes can lead to land degradation if it is to reduce the potential for land production. Most land degradation processes occur mainly in arid and semi-arid areas and cause adverse effects on land resources. Dryland has a high vulnerability to change and destruction, due to climatic pressures and the effects of population growth. Today, recognizing these changes is possible by comparing the differences created in a specific region of the images taken in several specified periods. The details of these changes are understandable by mapping the components of land resources into more than one period. Process Detection is a process that allows observing and recognizing differences in time series of phenomena, faults, and patterns of the earth's surface.

Chapter 2 - This study deals with agricultural land use in Ukraine. Providing general characteristics of processes in the Ukrainian agrarian sector witnessed over the past two decades, the authors estimate long-term trends and recent changes in agricultural land use and describe their drivers.

The analysis focuses on such a main tendency in land use change as agricultural land concentrating resulted in emerging large-scale agrarian holdings. The study aims to both test if the agricultural land concentration in Ukraine has a lot in common with the processes known throughout the world as "land grabbing" and substantiate priorities for countering its negative effects. The approach includes land policy analysis and evaluation, and case study analysis.

The analysis confirms that monopolizing control over agricultural lands by large agrarian holdings decrease the viability of both the agricultural sector and rural areas. Exclusion of agricultural land from local and rural development, illegal shadow" land market transactions and failing to meet the principles of responsible investments are the main features of land use by large corporate holders. Describing factors and consequences of land use concentration, the authors argue that many characteristics of such processes are evidence of land grabbing.

As the tendency to concentration of agricultural land become common almost worldwide, authors compare driving forces of agricultural land concentration, land grabbing, and their social consequences for rural areas in Ukraine and the EU countries.

Based on national features of changes in agricultural land use, marked by extensive land grabbing, the authors provide a rationale for stronger policy actions for both reducing the negative effect of large-scale corporate land use and supporting family farming that is characterized by more sustainable, socially and ecologically responsible land use.

Chapter 3 – The authors developed a framework which addresses sustainable land management through ecosystem services provided by healthy soils. The authors' conceptual tool is designed to support policy makers to manage five selected soil functions and demands: productivity, water availability, nutrients, carbon sequestration, biodiversity. The authors applied it to Amazon's Agricultural frontier - the Upper Xingu River Basin, state of Mato Grosso, Brazil – one of the most rapidly changing regions in the basin. In this study the authors implemented an interdisciplinary, GIS-based, multi-model approach to understand how land use change due to agricultural expansion and intensification is affecting soil functions at the Amazon's Arc of Deforestation. Encompassing two large Brazilian Biomes, the tropical rain forest and the Cerrado (Brazilian Savannas), the study area of ~170.000km^2 has undergone extensive changes in land use and land cover since the late 1970s. In only 40 years, the basin has already lost 30% of its natural vegetation, which was replaced mainly by pastures for cattle ranching. Since the early 2000s, a new cycle has started and currently the Upper Xingu River Basin is undergoing an increased agriculture intensification process (e.g., double cropping) to produce corn and soy bean for the international market. According to stakeholders, the main limiting factor for agriculture is infrastructure and absence of government. The lack of support affects markets and international trading with high costs for stocking and distribution of soy bean products. As proxies for soil indicator for supply the authors used a 2015 land use map derived from remote sensing data. Demand was expressed as annual productivity (kg/ha/y) from census data spatialized by land use type. Evapotranspiration was used as the proxy for water availability and water yield for demands. For a biodiversity indicator the authors used Indigenous Land and Conservation Units as suppliers and Legal Reserve and Permanent Protection Areas as demand (legal instruments). Nutrient Cycling supply was mapped based on a base saturation of 0-30 cm of almost 400 soil profiles and demand was mapped using average fertilizer application (NPK) spatialized by land use. Carbon storage and sequestration were derived from field measurements spatialized by land use and demands were generated using field carbon sequestration measurements and policy (Zero illegal deforestation target). The authors' results show that this approach can be applied to a range of landscapes and is a useful tool for decision making and policy implementation and support.

Chapter 4 - The implementation of seasonal forecasts of soil water in relative small spatial scales is of great interest, especially in the agricultural sector as they facilitate decision-making what allows a better management of water resources and maximize efficiency in productivity. In this work, Tres Arroyos meteorological station was chosen for the analysis. Located in the south of Buenos Aires, Tres Arroyos is one of the most important regions for corn production. The aim of this study is to design a statistical model using atmospheric forcing to predict soil water storage (WS) for spring. The analysis of the efficiency of different models takes into account the adjusted squared correlation coefficient ($\overline{R^2}$) and cross-validation coefficient (CV) values. The preliminary results show that the best designed model has an efficiency of around 66%.

In: Land Use Changes   ISBN: 978-1-53617-032-0
Editor: Vinícius Santos Alves   © 2020 Nova Science Publishers, Inc.

Chapter 1

APPLICATIONS OF LAND USE CHANGE PREDICTION MODELS (WITH REMOTE SENSING AND GEOGRAPHIC INFORMATION SYSTEM APPROACH)

Jafar Jafarzadeh[*]

Geography and Environmental Planning, Tabriz University,
Tabriz, Iran

ABSTRACT

Land use is one of the most important aspects of the study of natural resources management and environmental change review. Land use refers to human methods and purposes for resources. Identifying the timely and accurate changes in land use of the founder is a better way to understand the interactions of human and land resources. Recognizing these relationships leads to the management and sustainable use of these resources. Coverage and land use have two different meanings. Climate change, deforestation, flood and sediment increase, pollution, urban

---

[*] Corresponding Author's Email: jjafar1364@gmail.com.

growth, desertification, soil erosion and... is all of the consequences of development without planning and regardless of the environmental impacts of land-use change. The pattern of coverage and land use of an area, the output of the relationship between nature and socioeconomic factors, their application by humans, in the dimension of time and place. Modified land is a land that has changed in terms of climate, topography, soil characteristics and land use concerning their previous condition. The process of these changes can lead to land degradation if it is to reduce the potential for land production. Most land degradation processes occur mainly in arid and semi-arid areas and cause adverse effects on land resources. Dryland has a high vulnerability to change and destruction, due to climatic pressures and the effects of population growth. Today, recognizing these changes is possible by comparing the differences created in a specific region of the images taken in several specified periods. The details of these changes are understandable by mapping the components of land resources into more than one period. Process Detection is a process that allows observing and recognizing differences in time series of phenomena, faults, and patterns of the earth's surface.

Keywords: land-use, LULC, change detection, remote sensing

INTRODUCTION

Today, the destruction of soil and water resources is one of the main concerns of planners and managers in different parts of the world. Land degradation is a global process that ultimately leads to a decline in soil fertility and has become one of the major environmental issues around the world. This is due to pressure from population growth on land-based resources as a major challenge to food security and the quality of life for future generations, especially in developing countries such as Iran. Sustainable land management by preventing soil and land degradation is a factor in stabilizing and ensuring sustainable production for future generations and seems to be the only possible solution to the problem of natural resource degradation. In dry areas, land degradation is accompanied by extreme physical-biological and socio-economic phenomena that may turn into irreversible phenomena such as environmental degradation. Desertification is caused by the process of land degradation due to several factors in arid and semi-humid regions. Desertification is a process that affects some plant, soil and human ecosystems. The complexity and growing development of dynamic phenomena such as land degradation and desertification in the current century has focused on the use of new technologies for their evaluation and monitoring. The most important of these technologies based on spatial information technologies (geoinformatics) can be the remote sensing and geographic information system and global positioning system. The assessment of land degradation has been studied in several ways, and the best approach is the combination of remote sensing, geographic information, and field studies. Revealing changes is one of the basic needs in managing and evaluating natural resources. A Comparison of remote sensing and large-scale geographic information systems with conventional methods has shown that it is more cost-effective in terms of time and cost. Several studies have shown that the combination of remote sensing, geographic information systems, and field studies are an ideal and suitable method for identifying and classifying degraded areas of the earth.

The Capabilities and Applications of Satellite Remote Sensing

Remote sensing is used in all sciences that are in some way related to location information. Satellite data is widely used in agriculture and natural resources. Also, quantitative and qualitative conditions of agricultural products, recognition of certain types of products, tree identification, estimation of cultivars, development of products and production, pests and diseases are applicable. The study of forests and rangelands and their differentiation based on density, forest and pasture species, determining the role of salinity, water deficit, identifying halophytes and preparing land use maps is another of the uses of satellite data (land use refers to the specific use that humans use on land, For example, forest land, farms, irrigated land, dry land are examples of land use). One of the most important applications of the measurement data is to avoid the study of phenomena that are dynamic and changing over time. Dynamic phenomena in agriculture and natural resources include the growth of crops, soil degradation and vegetation, land degradation and desertification. Because desertification and destruction of the land occur over time, satellite data can be used to estimate the location and time of desert land. Remote sensing data due to its integrated and wide variety, spectrum diversity, duplication, and low-cost coatings, has a special feature in comparison with other information-gathering methods, which today is the primary factor in studying the surface of the earth and its constituent factors. Data digitization has made it possible for computer systems to use these data directly, and geographic data systems and satellite data processing systems are designed and developed using this capability. Easy access to data, quick access to remote sites and high accuracy are the special advantages of this fan. Satellite data can play an effective role in providing land cover mapping and land use maps due to its specific features including wide coverage, repeatability, multi-spectral, diversity. The successful use of remote sensing techniques in these studies is deeply dependent on factors such as a sufficient understanding of the landscape of the region, the type of sensor, and the method of extracting information. The most important environmental factors in these evaluations include the study of atmospheric conditions, soil moisture and phonological properties of plants at the time of image capture. Of course, the challenge posed by the use of this technology is to ensure short or long term land-use change.

Applications of land use change prediction models are:

- prediction of changes in the growth and development of cities in the future
- Forecasting forest changes (deforestation) in the future and presenting the best scenario for preventing deforestation.
- Analysis of changes in terrestrial and landscape ecology in the next years
- Runoff modeling, erosion and sedimentation using exit models of land use change forecasting models
- Modeling of habitat and biodiversity evaluation using exit models of land use change forecasting models

- Planning to choose the best corridors and crossings between wildlife habitats in the future
- Assessing the sensitivity of wildlife habitat to landscape changes in the coming years
- Surveying and spatial analysis of lake changes (e.g., Urmia Lake)
- Estimation of greenhouse gas emissions, especially carbon dioxide, using the output of land-use change prediction models

In this section, more comprehensive descriptions of applications of land-use change prediction models are presented:

Prediction of Changes in the Growth and Development of Cities in the Future

The growth of cities around the world has led to the destruction of terrestrial resources and the destruction of vast areas of pristine and natural lands and their transformation into impenetrable levels. In many cases, these changes have occurred without understanding their implications. To analyze the land in cities, models can be used as a means to guide urban design for sustainable development. Using the methods of detecting changes and modeling, one can analyze past and predict future city growth. The growth of cities has affected natural environments and changed the structure and role of ecosystems. Today, cities are developing and expanding at very high growth rates. This issue of urban growth has become one of the most important challenges of the 21st century. Meanwhile, providing spatially-temporal data related to the pattern and the city's growth rate is essential for a better understanding of the urban growth process and the adoption of very important management policies. Although cities occupy only 3% of the world's surface [16], they are growing at an unpredictable rate and scale all over the world [17], especially in developing countries [8].

Urban dispersal development refers to the outdated and unplanned development of cities, which not only damage the surrounding space but also cause uneven urban development. This phenomenon has several effects on local, regional and global scales. These effects include increasing pollution of water resources, soil and air, increasing energy consumption and intensifying the phenomenon of islands of urban heating, climate change, degradation of vegetation, agricultural lands and also, the negative effects on the mental and physical health of the residents of the cities[14, 18].

Increasing the population, in turn, leads to an unplanned physical and unplanned development, increasing marginalization and creating settlements around large metropolises and cities. Under these conditions, physical development usually occurs without regard to the natural and biological parameters. The destruction of gardens and agricultural lands in favor of construction, the achievement of environmental values, the development of steep slopes, inappropriate neighborhoods in applications, including the consequences of this kind of physical development (Figure 1).

Figure 1. The expansion of the city of Riyadh, the capital of Saudi Arabia (left photograph in 1972, middle photo in 1990 and right photo in 2000).

Analysis of Changes in Terrestrial and Landscape Ecology in the Next Years

Over the past years, humans have changed ecosystems (Figure 2) more quickly and more exponentially than any period in human history[4]. So that human intervention in the environment has led to a change in the natural ecosystem, environmental degradation, and the transformation of natural attractions into residential and industrial uses that have driven humanity away from nature.

Urban Land Cover 1990-2020

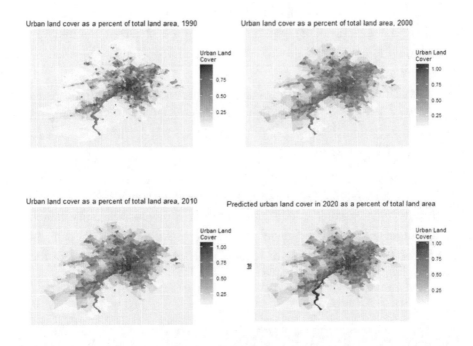

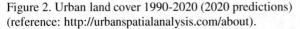

Figure 2. Urban land cover 1990-2020 (2020 predictions)
(reference: http://urbanspatialanalysis.com/about).

One of the issues of the day is the issue of environmental protection. Increasing global environmental problems and crises on the one hand, and understanding the long-term consequences of environmental issues on the other, has increased the importance of discussing the environmental issues over the last half century [2].

To know the status of land changes, we need to examine the changes in the spatial and temporal scales. Territory landmarks, metrics such as the number of spots, the density of stains, the margin density, the Shannon variety, the measure of the largest spot, ... to quantify the trend and pattern of landform changes to make changes in a landscape over time or a comparison between perspectives.

Also, the diversity and diversity of these measures have led to their widespread use in land-related planning. Landscape patterns can be analyzed with the Arc GIS software combined with other software and models such as Flag state and Land Change Modeler (LCM). When the land use map was developed with land-use change (Figure 3) prediction models, the trend of land map changes from the present to the future can be quantitatively calculated with metrics and provide good planning for land management in the coming years.

Natural tourism in the current world is pure industrial and the third is the dynamic economic and developmental phenomenon, which after the oil and automobile industry has been abducted from other global industries. Obviously, the greater the number of tourist attractions in the area, the greater the demand for visitors and the higher the number of visitors, the greater the degradation and vulnerability of tourist attractions.

Because each region has a different ecological tolerance capacity that is: the maximum use that can be made from each area without such use causing negative effects in the resource or reducing the visitor's satisfaction, or the effects it undermines the community, economy, and culture of that region[15].

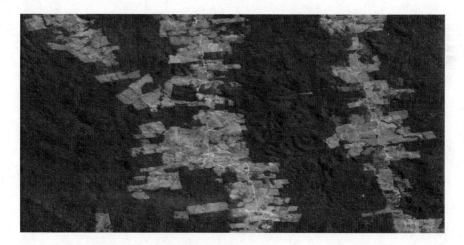

Figure 3. Detect deforestation and changes in land use
(reference: https://www.planet.com/markets/forestry/).

Runoff Modeling, Erosion and Sediment Modeling Using Land-Use Model Prediction Models

Land-use changes, especially the growth of urban areas, have led to an increase in sediment in many parts of the world. Land-use changes also affect the cycling of nitrites, hydrology, and climate. Exit map of land use change prediction models can be entered as input into sediment simulation and hydrology model, and estimate sediment and runoff levels quantitatively in the coming years. Erosion, sediment load, runoff, and evapotranspiration - transboundary watersheds and rivers for optimal use of water resources are one of the most important challenges facing the country's water resources management, which are effective in exploiting water facilities and dams. Soil erosion is one of the most serious environmental issues in the world. Given the high rates of general erosion in many parts of the world, much effort must be made to reduce its risks. This requires little data to identify and identify critical areas that require immediate protection. Traditional methods of doing these things are first and foremost expensive, secondly, they get point data. Therefore, it is necessary to use the new technologies of remote sensing and GIS to conduct systematic research. The sedimentation in the dam reservoir not only reduces the capacity and useful life of the reservoir but also causes many problems in opening and closing the deep and semi-deep openings of the dam, increasing the reservoir's surface and increasing the evaporation and water losses. These issues highlight the importance of investigating the phenomenon of sediment accumulation in reservoirs to predict the distribution of sediment and provide management solutions for reservoir control. Mathematical models are one of the most important tools for predicting the amount of sedimentation in reservoirs of reservoirs and estimating their useful life. These models are based on the analysis of the governing equations on the phenomena affecting the transfer, distribution, accumulation, and dehumidification of sediment. Today, remote sensing and geographic information systems of advanced systems of natural resources studies, especially in soil erosion studies and sedimentation of watersheds come. Geographic information systems (Figure

4) are computer systems that used to collect, store and use geographic information.

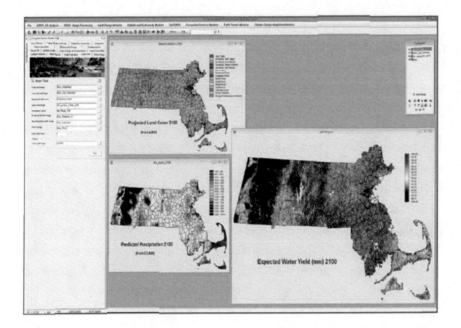

Figure 4. The bottom right figure shows the annual runoff estimated in 2100. The input map for land use project in the year 2100 (upper left figure) was produced with CCAM to produce this image using the Land Change Modeler and the Precipitation Forecast (bottom left figure) of 2100.

Habitat and Biodiversity Assessment Modeling Using Exit Models of Land Use Change Forecasting Models

Predicting and assessing the potential of land-use patterns through modeling can help environmental planners and natural resource managers to make informed decisions. Human use of the earth changes the structure and function of the ecosystem. The most important spatial and economic use of human beings from the earth around the world includes various cultivations, structures, reserves, protected lands and minerals extraction[1,19]. The importance of vegetation or land-use as a dynamic and effective factor on biological (Figure 5) conditions requires that accurate quantitative and qualitative information should always be provided and its changes in short-term intervals are determined[9]. Using RS and GIS techniques, preparing a land-use map at different times and comparing them with each other to achieve the map of the changes made easily can be done. After recognizing the changes made during the period, the relationship between the changes and the factors involved is studied[7]. Using the multi-sensory data of remote sensing with the least time and cost can be used to extract land use, and then, by comparing it in different periods, the ratio of changes is evaluated [13]. Many land-use change modeling studies are carried out using time series remote sensing imagery to extract the effective parameters for user-oriented transformation across the region [3].

Figure 5. Biodiversity Logo. https://www.pinterest.com/pin/154177987219838223/.

Planning to Choose the Best Passageway between Wildlife Habitats in the Future

The maps drawn in this section can be used to accommodate the maximum population of wildlife in the ecosystem and as a primary source of planning for future corridors. It is also effective in determining which areas can serve as potential crossings. The streets and the roads are communication paths that connect and link two places or in the context of the community of human beings. In all definitions of the road, there is the concept of linking, but in the history of centuries of building roads (Figure 6), as much as the link between spaces, there is also a disconnect. By identifying the destructive role of roads and highways in the ecological rupture of landscapes, it is time to review roads and streams and re-examine the main role of the road as a linker, not the elements that blow up natural and animal habitats. Today, designers and planners are expected to consider various ways of designing and constructing them in addition to choosing the best place to build intercity roads and highways. Roads and highways should be the elements that create the link between humans and wildlife, culture and nature. Designing and planning the landscape is one of the most important tasks of landscape architects. The present text examines and analyzes the examples and ideas presented for the ecological corridors in the arc competition, and the effect of these components on the design and planning of the road perspective and their role in creating the continuity of the landscape and ecology of the road.

There are 28 million kilometers of highways in the world. Over the past sixty years, the number of cars in the United States has more than tripled. The development of cities and the unceasing growth of human settlements and, consequently, the need to build highways and roads, in many cases, have eroded the natural landscape and animal habitats (Figure 7). Roads and highways can have a profound negative impact on the sustainability of the natural landscape, and thus on the sustainability of wildlife and the ecological transplantation of the landscape. Roads can cause animal species to lose their habitat, and in addition prevent the movement of a wide range of organisms, including invertebrates and small and large mammals. Even elementary and simpler streets can have important negative impacts on the habitat's survival of some animal species. Building roads and highways will eliminate plant species and animal deaths. Many animals also lose their lives while traveling through roads. Animal health statistics have doubled in the United States over the last fifteen years. The main problem is when roads cut off routine routes and migration routes of animal populations, and the animals are taken up by cars through their usual paths.

Figure 6. Canadian Highway Canada Transit and Pacific Rail Canada are important transit routes that cross one of the most important Rocky Hills.[The source: www. Humans and nature.org/reweaving].

Figure 7. Dealing with deer is the most common accidents related to vehicle savvy, cars, and road safety in the United States. [The source: http://www. cultureofsafety.com/driving/deer-vs-car-collisions].

Assessment of the Sensitivity of the Wildlife Habitat to Changes in the Landscape in the Coming Years

The basis for the protection of biodiversity is protected marine and land conservation areas. These areas include national parks, protected landmarks and countless storehouses [10]. Protected areas include conservation of biodiversity, the preservation of cultural heritage, the maintenance of vital ecosystem services and socio-economic (Figure 8) benefits [11]. By investing in money, land, and labor, the community seeks to use, manage and preserve the role of protected areas.

On the other hand, in the second half of the last century, most countries have become the core of the strategy for the protection of biodiversity and the environment, protected areas they have increased their numbers and sizes [10]. The habitat quality and rarity model examines the effects of human threats on the quality and rarity of habitats, one of the inputs of this model, a map of the future land use prediction. This is a general model for assessing the sensitivity of the habitat due to changes in the landscape. The results of this model are used for quick evaluation of regional habitats, which can be used as a representative for deep insight into the species adaptation status. The model maps the quality of habitat and rarity that represents the change of habitat over time.

Today, due to the lack of logical use of land, the conversion, change in coverage and land use in the territories have been growing. Reviewing and quantifying these changes can be a disaster in the planning and management of sustainable land.

Land Use /Land Cover (LULC) as a result of complex interactions the structural and functional factors associated with demand, technological capacity and social communication have a widespread impact on the land's shape [12]. In recent years, many studies have benefited from the land-based landmarks for land planning and management, most of which have been using time series measurements to study land cover changes, especially in the United States and China [20].

Figure 8. Wildlife_corridor & client. [Ref: https://upload.wikimedia.org/wikipedia/commons/thumb/3/36/Bobcat_urbanLandscape.jpg/440px-Bobcat_urbanLandscape.jpg].

Surveying and Spatial Analysis of Lake Changes (E.G., Urmia Lake)

Investigating the fluctuations in lake water levels in recent years has been important in terms of the importance, nature, and position of these water complexes. Investigating the local variations of water pollution is very important in the management of water resources crisis and the prediction of climate change. Water quality and its impact on human health and living organisms are always considered as one of the major challenges in underdeveloped or developing societies. Today, in each national economy, the dynamics of regional development programs are among the key pillars of growth and development. The need for any logical planning of knowledge is from the natural and human environment. Although the natural environment itself is a self-regulating device, the breadth of human performance without planning and applying proper management has led to natural changes.

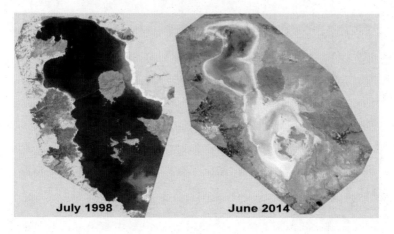

Figure 9. The Landsat satellite image of Lake Urmia for 1998 and 2014.

Lake Urmia (Figure 9) is one of the largest lakes in the I.R of Iran and the world's super saline lakes in the world. It is one of the international wetlands under the Ramsar convention, which is of great importance in the economic, social, tourism and environmental sectors of this region of Iran. The recent environmental crisis and drying of Lake Urmia are one of the greatest environmental hazards in Iran. The downward trend in the balance over the past years has been a source of serious concern. On the other hand, the growth of cities throughout the world, especially developing countries, has led to the elimination of land resources and the destruction of vast areas of water and the conversion of them to impenetrable levels. In many cases, these changes have occurred without understanding their implications. In this context, monitoring and evaluating such areas can be considered as an important factor in national development and management of natural resources. In this regard, remote sensing technology has taken on a unique role in obtaining information from these phenomena, because multi-spectrum satellite imagery has advantages and privileges that their accessibility and digital divisions are among the most important attributes of this phenomenon. Gets Models can be used as an appropriate tool for managing management for sustainable development to analyze natural resources, especially blue. Using change detection and modeling techniques, we can show the trend of past change and predict future growth. Land-use change models are land-use management and sensitive zones and can identify land-cover changes in the future according to different scenarios.

The Aral Sea was an endorheic lake lying between Kazakhstan (Aktobe and Kyzylorda Regions) in the north and Uzbekistan (Karakalpakstan autonomous region) in the south. Formerly the fourth largest lake in the world with an area of 68,000 km² (26,300 sq mi), the Aral Sea has been shrinking since the 1960s after the rivers that fed it were diverted by Soviet irrigation projects. By 1997, it had declined to 10% of its original size, splitting into four lakes: the North Aral Sea, the eastern and western basins of the once far larger South Aral Sea, and one smaller intermediate lake. By 2009, the southeastern lake had disappeared and the southwestern lake had retreated to a thin strip at the western edge of the former southern sea; in subsequent years, occasional water flows have led to the southeastern lake sometimes being replenished to a small degree. Satellite images taken by NASA in August 2014 revealed that for the first time in modern history the eastern basin of the Aral Sea had completely dried up. The eastern basin is now called the Aralkum Desert.

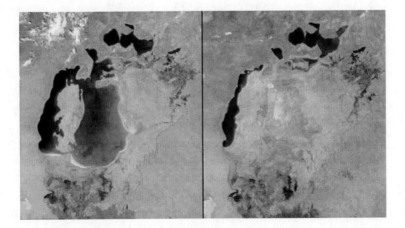

Figure 10. Destruction and drying of the Aral Lake in Kazakhstan (left photograph in 1960 and right pic in 2000).

Estimation of Greenhouse Gas Emissions, Especially Carbon Dioxide, Using the Output of Land Use Change Prediction Models

During the past century, the rapid growth of technological advances in industrial societies has significantly increased the contribution of human intervention to climate and temperature changes caused by greenhouse gases are one of the scientific and political challenges of the present era. Since the emphasis on environmental protection has been highlighted in all international and national laws, humans have to work hard to maintain their lives.

All living organisms of the world are an integrated system in which every shock in each region affects the whole system. Hence, the environmental hazards and challenges are not an internal issue, but a global issue. The importance of this issue concerning air pollution is more than any other factor. Since there is no boundary in the atmosphere, emissions of greenhouse gases in a given place and time can, in time, appear in another place. Many of the pollutants from fossil fuels released in the atmosphere around the globe raise the concentration of greenhouse gases and, as a result, increase the global warming of the planet. The latest information from the world's scientific centers shows that the rate of greenhouse gas emissions such as carbon dioxide, methane in the earth's atmosphere increased by 158% and 19%, from 1785 to 1850, respectively [IPCC, 2007].

However, the role of human activities in increasing greenhouse gas emissions has been proven as responsible for climate change and global warming. Scientific reports show that greenhouse gas concentrations in the atmosphere have increased by about 35°C during the last decades and temperatures of about 1°C [IPCC, 2009].

Considering the importance of global warming (Figure 11) and the rising trend of rising ground temperatures during recent decades and the destructive effects of this phenomenon on various sectors of the economy and the environment, it is imperative to model the factors affecting the increase of average ground temperature in the future in order to control and or eliminating and reducing its causes is more than ever known. Land use changes play an important role in the global carbon cycle, which has produced about 136 gigs of carbon since the start of the industrial revolution as a result of land-use changes to the atmosphere. The United Nations Convention on Climate Change (UNFCCC[1]) endorses forests as a source of greenhouse gas savings and their vital role in the global carbon cycle. The role of forests in reducing climate change has been addressed at the Greenhouse Gas Emissions Reduction and Disposal (REDDF[2]) agenda. The main goal of REDD is to reduce greenhouse gas emissions, but it also includes benefits such as biodiversity protection and poverty reduction. Remote sensing, land-use change detection, land-based land plot maps, and land-use change models are key factors in forest loss, forest management, and carbon reservoirs.

Figure 11. Global Warming. (Ref: https://theconversation.com/what-would-happen-to-the-climate-if-we-stopped-emitting-greenhouse-gases-today-35011).

[1]. United Nations Framework Convention on Climate Change.
[2]. Reducing emissions from deforestation and forest degradation.

CONCLUSION

One of the environmental hazards and ecological crises that the world now faces is the phenomenon of land use change. Land-use is a special use of humans from the earth. These uses are changing over time and these changes lead to increased destruction of the land and the destruction of the ecosystem, especially in arid and semi-arid regions. Therefore, to control and combat the crisis, user changes require the recognition and understanding of the underlying processes and its future trend. Urgent growth of cities and increased pollution of resources, the destruction of a large area of forests, erosion of agricultural lands, the occurrence of destructive floods, and the spread of deserts ecosystems are often due to the unconventional land cover and the use of inappropriate methods of exploitation is used.

Today, remote sensing is widely used to monitor changes in land cover and land use and their dynamics. Using satellite imagery and digital processing with appropriate algorithms can minimize human error, detect and differentiate the cost and time of the phenomenon details. To achieve better results in identifying the event changes, remote sensing is often combined with geographic information systems. There are several methods for extracting land cover maps, in which the remote sensing technique (Figure 12) is important because of certain characteristics. Planning and proper management of the land require accurate and up-to-date maps. Different techniques of remote sensing by generating maps of land cover and land use provide planners with the necessary information and help them in decision making.

The increasing acceptance of remote sensing methods by researchers and decision-makers at different levels confirms the appropriateness of this technology in such studies. Therefore, it is essential to review the concepts of land degradation and desertification and its importance and important place in the sustainable development of natural resources, so that their methods of evaluation and monitoring can be considered to be able to introduce a more appropriate method.

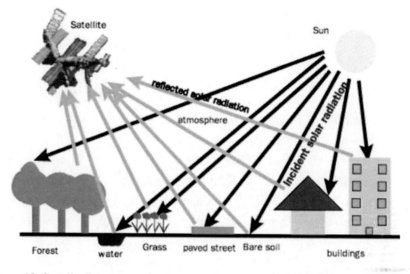

Figure 12. Sattelite Remote sensing.

REFERENCES

[1]   Awotwi, A. 2009. *Detection of land use and land cover change in Accra, Ghana, between 1985 and 2003 using Landsat imagery.* MSc. Thesis, Division of Geoinformatics Royal Institute of Technology (KTH), Stockholm, Sweden.

[2]   Budak, D. (2005). "Behavior Attitude of student Toward Environmental Issues at faculty of Agricultural." Turkey, *Jurnal of Applied sciences,* 1224-1227.

[3]   Conway, T. M. and Lathrop R. G. Jr. 2005. Modeling the ecological consequences of land-use policies in an urbanizing region. *Environmental Management,* 35:278-291.

[4]   Fennel, D. A. (2006). "*Introduction on Ecotourism.*" Translated Chadiko laei. Babolsar, Mazandaran University Press. 134.

[5]   Gumeh, Z. 2012. *Monitoring land use changes in Karaj and its association to landscape profiles,* M. Sc. thesis, University of Shahid Chamran, Ahvaz, 128p.

[6]   IPCC (2007) Climate Change 2007: *"The Physical Science Basis. Contribution of Working Group I to the Fourth Assessment Report of the Intergovernmental Panel on Climate Change,"* S. Solomon, D. Qin, M. Manning, Z. Chen, M.

[7]   IPCC (2009) Climate Change 2009: *"The Physical Science Basis. Contribution of Working Group I to the Fourth Assessment Report of the Intergovernmental Panel on Climate Change,"* Cruz, R. V. O., R. D. Lasco, J. M. Pulhin, F. B. Pulhin and K. B. Garcia: Climate change impact on water resources in Pantabangan Watershed, Philippines.

[8]   Karolien, V., V. R. Anton, L. Maarten, S. Eria, and M. Paul. 2012. Urban growth of Kampala, Uganda: Pattern analysis and scenario development. *Landscape and Urban Planning* 106: 199-206.

[9]   Koomen, E., Stillwell, J., Bakema, A., and Schol ten, H. J. 2007. *Modeling Land-Use Change, Progress and Applications.* Springer, Dordrecht, the Netherlands, 390p.

[10] Leverington, F., Costa, K. L., Courrau, J., Pavese, H., Nolte, C., Marr, M., et al. 2010. Management effectiveness evaluation in protected areas – *A global study.* 87. https://doi.org/10.1007/s00267-010-9564-5.

[11] Li, Y., Sun, X., Zhu, X., & Cao, H., 2010. An early warning method of landscape ecological security in rapid urbanizing coastal areas and its application in Xiamen, China. *Ecological Modelling,* 221(19), 2251–2260. https://doi.org/10.1016/j.ecolmodel. 2010.04.016.

[12] Matsushita, B., Xu, M. and Fukushima, T., 2006. Characterizing the changes in landscape structure in the Lake Kasumigaura Basin, Japan using a high-quality GIS dataset. *Landscape and urban planning,* 78(3), pp.241-250.

[13] Rabiee, H., Ziaeean, P. and Alimohammadi, A. 2004. Exploring land uses and land cover in Isfahan province using remote sensing and GIS. *Journal of Geographical Research,* 84:41-54 (In Persian).

[14] Rafiee, R., A. Salman Mahiny and N. Khorasani. 2009. Assessment of changes in urban green spaces of Mashad city using satellite data. *International Journal of Applied Earth Observation and Geoinformation* 11: 431-438.

[15] Rezvani, M. Nojavan, M. Bakhtyary Roodsari, A. (2014). "Evaluation of the effects of environmental Bragahyhay informal Tour ecotourism." *Quarterly Journal of Environmental Education and Sustainable Development.* 1(1), 25-45. [In Persian].

[16] Sabet Sarvestani, M., A. Latifi Ibrahim, and P. Kanaroglou. 2011. Three decades of urban growth in the city of Shiraz, Iran: A remote sensing and geographic information systems application. *Cities* 28(4): 320-329.

[17] Sun, C., Z. Wu, Z., Lv, N. Yao and J. Wei. 2013. Quantifying different types of urban growth and the change dynamic in Guangzhou using multi-temporal remote sensing data. *International Journal of Applied Earth Observation and Geoinformation* 21: 409-417.

[18] Thapa, R. B. and Y. Murayama. 2012. Scenario-based urban growth allocation in Kathmandu Valley, Nepal. *Landscape and Urban Planning* 105(1-2): 140-148.

[19] The, J. 2006. Detection of changes using remote sensing: an overview of principles and applications. *Geo-Spatial and Range Sciences Conference.* Idaho State Univ., Pocatello, ID, USA.

[20] Uuemaa, E., Mander, Ü. and Marja, R., 2013. Trends in the use of landscape spatial metrics as landscape indicators: a review. *Ecological Indicators*, 28, pp.100-106.

In: Land Use Changes ISBN: 978-1-53617-032-0
Editor: Vinícius Santos Alves © 2020 Nova Science Publishers, Inc.

Chapter 2

LAND CONCENTRATION OR LAND GRABBING: MULTIDIMENSIONAL ISSUES CHALLENGING UKRAINIAN AGRARIAN SECTOR

Olena Borodina and *Viktor Yarovyi*[†]

Department of Economy and Policy of Agrarian Transformations,
Institute for Economics and Forecasting,
National Academy of Sciences of Ukraine, Kyiv, Ukraine

ABSTRACT

This study deals with agricultural land use in Ukraine. Providing general characteristics of processes in the Ukrainian agrarian sector witnessed over the past two decades, the authors estimate long-term trends and recent changes in agricultural land use and describe their drivers.

The analysis focuses on such a main tendency in land use change as agricultural land concentrating resulted in emerging large-scale agrarian

---

[*] Corresponding Author's Email: olena.borodina@gmail.com.
[†] Corresponding Author's Email: v.yarovyi@gmail.com.

holdings. The study aims to both test if the agricultural land concentration in Ukraine has a lot in common with the processes known throughout the world as "land grabbing" and substantiate priorities for countering its negative effects. The approach includes land policy analysis and evaluation, and case study analysis.

The analysis confirms that monopolizing control over agricultural lands by large agrarian holdings decrease the viability of both the agricultural sector and rural areas. Exclusion of agricultural land from local and rural development, illegal shadow" land market transactions and failing to meet the principles of responsible investments are the main features of land use by large corporate holders. Describing factors and consequences of land use concentration, the authors argue that many characteristics of such processes are evidence of land grabbing.

As the tendency to concentration of agricultural land become common almost worldwide, authors compare driving forces of agricultural land concentration, land grabbing, and their social consequences for rural areas in Ukraine and the EU countries.

Based on national features of changes in agricultural land use, marked by extensive land grabbing, the authors provide a rationale for stronger policy actions for both reducing the negative effect of large-scale corporate land use and supporting family farming that is characterized by more sustainable, socially and ecologically responsible land use.

Keywords: agricultural land use, land concentration, land grabbing, land raiding, agricultural policy, large-scale farming, family farming, corporatization of agriculture.

INTRODUCTION

We start with a description of the background that made possible the excessive land concentration and land grabbing in Ukrainian agriculture.

As of 2017, Ukraine had 41.5 million hectares of agricultural land, including 31.1 million hectares of private land and 10.4 million hectares of state-owned land. Individuals own about three quarters of agricultural land, including 6.9 million owners of 27.7 million hectares of so called "land shares" received by former workers of collective farms in the 1990s.

Ukrainian agriculture consists of corporate and individual sectors. The corporate sector includes approximately 14 thousand big corporate farms.

The individual sector includes more than 34 thousand peasant farms and 4 million individual farms (farming households). Peasant farms are legal entities, but individual farms are households operating on their land usually without legal status and producing agricultural goods for both self-consumption and market, but mostly for self-consumption.

Corporate and individual farms are the main users of agricultural land in Ukraine. Corporate agricultural enterprises get land from both small private owners and state through leases, and, to maximize profit, focus on producing commercial crops with high export potential. It stimulates corporate farms to enlarge their land banks leading to land concentration and land grabbing.

Peasant farms that are legal entities use their own land, land shares leased from households and leased state land (about 0.75 million rental agreements as of 2018). Some peasant farms, which are legal entities, are real family farms. There are also many that are more similar to big corporate farms than to small family farms because of large land banks and a lot of hired employees.

The most part of the land owned by farming households is used by corporate farms under lease agreements – about 16.9 million hectares or 4.9 million agreements as of 2018. The main reasons why individual farming households do not establish their own commercial farms are limited access to financial and other resources, and institutional environment that is not conducive to improving the competitiveness of small family farms.

Most of the state-owned agricultural land should be transferred to local communities. Until then, the State Service of Ukraine for Geodesy, Cartography and Cadastre (StateGeoCadastre) is the administrating agency for most state agricultural land, and such land mostly is used by corporate farms under lease agreements. In 2018, Ukrainian government slowly started transferring state lands to communal property of local communities, but this process is still far from complete as of August 2019.

To understand the processes of land concentration and land grabbing in Ukraine better, it is also necessary to mention the political context, especially disputes on the agricultural land market and the moratorium on selling agricultural land. As at mid-August 2019, the official agricultural land market in Ukraine is represented mostly by land lease transactions, and the moratorium for selling and buying agricultural land is in force. Many policymakers believed that the moratorium was a contributing factor in land concentration in Ukraine, including the non-transparent and semi-legal practices that we marked as land grabbing.

In the first half of the 1990s, during President Leonid Kravchuk administration, nobody has even discussed the agricultural land market in Ukraine.

Up to the end of the 1990s, millions of former members of collective farms received land shares or rights to land plots. That created the preconditions for launching the agricultural land market. Considering the fact that the legislation for agricultural land market was absent, as well as the absence of land cadaster and land register, the Ukrainian Parliament adopted the Law on Disposal of Land Share (*Law of Ukraine "On Disposal Of Land Share"* 2001) and temporarily banned selling land shares. Later that year, the Parliament adopted the Land Code that temporally extended the ban for selling and buying land to all agricultural lands. Since then and to the mid-August 2019, the moratorium has been extended many times.

By the end of President Leonid Kuchma's second term in 2004, he believed that the extension of the moratorium would be beneficial for some large actors and holdings interested in dictating their own terms and requirements to small landowners. President Kuchma tried to lift the moratorium on purchasing land, but failed in his confrontation with the Ukrainian Parliament, and the moratorium was extended.

President Viktor Yuschenko believed that the moratorium was the primary reason for corruption in land relations. As of 2007, the moratorium resulted in the concentration about 39 percent of agricultural land area by new owners through non-transparent and semi-legal practices. In his opinion, because the moratorium violated the rights of small landowners and contributed to the increase of semi-legal practices, it should be lifted. President Yuschenkoa tried to lift the moratorium several times, but also failed like his predecessor because of strong opposition of the Ukrainian Parliament.

President Viktor Yanukovich also declared the intension to ban the moratorium on purchasing land during his administration. He succeeded in lobbing the Law on State Land Cadastre, which was adopted by Ukrainian Parliament. But lobbing the Law on Land Market, the second law that was necessary to adopt before lifting the moratorium, led to resistence in Ukrainian Parliament and popular protests, especially within ogranizations representing small farmers and farming households. At the same time, those actors close to the power continued to use non-transparent and semi-legal practices to take control over agricultural land. Also, establishing a new institution, State Food and Grain Corporation of Ukraine, led to consolidation about 1 million hectares of agricultural land.

President Petro Poroshenko also declared his support for abolishing the moratorium for sale and purchase of agricultural land, stating that uncertainties with land reform would lead to the situation in which only speculators and swindlers would benefit from the current state. But during the President Poroshenko's administration term (2014-2019), the moratorium was extended three times because Ukrainian Parliament failed to adopt the legislation that is necessary for lifting the moratorium.

New Ukrainian President Volodymyr Zelensky declared to lift the moratorium by the end of 2019 ("New Ukrainian President Wants to Allow the Sell of Agricultural Land by the End Of 2019" 2019).

Thus, political struggle shaped the permanent background for continuous Ukrainian land reform with many uncertainties and gaps in regulations of land relations resulted in agricultural land concentration and land grabbing as further described in this chapter.

This study is designed to answer the questions as follows: What are the drivers and extent of land concentration in Ukrainian agriculture? Are there any evidences of land concentration and land transactions to recognize them as land grabbing? If yes, how agricultural land grabbing would manifest itself in Ukraine? Who are the main actors forcing land concentration and land grabbing? What are the evidences of extreme land concentration and grabbing in Ukraine? What are the similarities and differences of land grabbing in Ukraine and, let us say, in EU-countries, where there are many land regulations to protect rural areas and small family farms? What are the real opportunities to deal with extreme land concentration and grabbing in the short term?

THEORETICAL PERSPECTIVES

The Formulation of Land Concentration

Land concentration in agriculture is a worldwide trend. According to the Global Network for the Right to Food and Nutrition, t*he degree of land concentration* refers to the structural repartition of agricultural holdings within a given territory, reflecting the extent of farmland controlled by small or large agricultural holdings (Global Network for the Right to Food and Nutrition 2019). To express the *degree of land concentration, the* Global Network for the Right to Food and Nutrition suggests indicators as follows: the percentage of land controlled by small-scale farmers and the Gini coefficient for land distribution (Global Network for the Right to Food and Nutrition 2019).

The Formulation of Land Grabbing

According to the definitions of Baker-Smith and Szocs-Boruss, land grabbing is the control (whether through ownership, lease, concession, contracts, quotas, or general power) of larger than locally typical amounts of land by any persons or entities (public or private, foreign or domestic) via any means ("legal" or "illegal") for purposes of speculation, extraction, resource control or commodification at the expense of agroecology, land stewardship, food sovereignty and human rights (Baker-Smith and Szocs-Boruss 2016, 15). European Coordination via Campesina, a European grassroots organization that currently gathers 31 national and regional rural organizations based in 21 European countries, also follows this definition, emphasizing that this control of land comes also at the expense of peasant farmers (European Coordination Via Campesina 2016).

When analyzing commons grabbing, Dell'Angelo J., D'Odorico P, Rulli M. C. and Marchand P. acknowledge coercion as a constitutive signal of the presence of land grabbing (Dell'Angelo et al. 2017).

Thus, land concentration and land grabbing may occur simultaneously. These processes are often interlinked. Nevertheless, not every land concentration has the characteristics of land grabbing. While land concentration may vary in nature, land grabbing always builds on a seizure of control over land for benefits. Land grabbing may lead to land concentration, but land concentration is not necessarily a consequence of land grabbing.

Land grabbing is a worldwide phenomenon, revealing itself in different ways depending on local specificities. However, there are at least two common features. The first one is the underlying process of capital accumulation. Land grabbing is a process initiated by actors who have the financial resources and political power, allowing access and control over agricultural land. The second one is that land grabbing is typically accompanied by the violations of human rights and the rights of current land users.

Usually, land grabbing is accompanied by the concentration of agricultural land in the hands of a very small proportion of the population. Land grabbing does not mean just changes in land ownership and land management. There are also deep changes in social, economic, institutional, and environmental dimensions of agricultural production.

In this study, we use the term "land grabbing" to refer to the processes of land seizure or gaining control over land use for agricultural production. At the same time, we are aware of the important role of all the other considerations and motives for land grabbing.

The Formulation of Land Raiding

Takeover of business in a competitive environment, which is often referred to as raiding, or corporate raiding, is a worldwide natural economic phenomenon. Raiding is usually strictly regulated at the legislative level. In the post-Soviet space, particularly in Ukraine, people perceive this phenomenon in a negative light, since it is a takeover of business against the will of its owner. In the communication No. 2027/2011 to the United Nations

Human Rights Committee, for instance, the concept of corporate raiding refers to the seizure by one party of an asset from and against the will of another party by means of threat, pressure or violence.

Raiding is a process that consists of both legal and illegal activities of the individuals or entities to seize or takeover business entities of any form of ownership leading to a full control over them. Whether legal or illegal actions are taken, raiding in Ukraine is always implemented in a way that constitutes an offence (Shapiro and Bulakh 2015). According to the terminology used by the Ministry of Internal Affairs of Ukraine, raiding is the forcible seizure of contentious entities, including cases of sentence enforcement in favour of any particular owners (Panasenko 2010).

The main actors involved in land raiding in Ukraine on the side of raiders are instigators and actual perpetrators. Instigators (natural or legal persons) are usually the sponsors. Some of the perpetrators are persons paid to work for large companies providing perpetrators with all resources that are necessary for land raiding.

There are also independent perpetrators operating unilaterally, using their own resources, and gaining all the revenue from the illegal of semi-legal land turnover. Land raiding in Ukraine is usually acquisition of temporary land rights with onward sale as soon as possible. Raiders usually sell the land rights to their own shell companies that sell the rights once again, and again to the affiliated companies. As a result, it is very hard for the original, legitimate owners to reclaim their land rights.

METHODS

We started by defining three terms: the land concentration, land grabbing, and land raiding. For this, we analysed the existing concepts of these terms, international and Ukrainian practice of using these terms, and defined their key characteristics that is most relevant to the Ukrainian circumstances and this study.

To analyze agricultural land use in Ukraine, we used the official statistical data on enterprises engaged in agricultural activity provided by

State Statistics Service of Ukraine. To define the degree of land concentration, we studied the distribution of farmland across agricultural holdings with different land area. We calculated and compared the percentages of land controlled by small-scale farms and largest farms. We followed the European approach, according to which small-scale farms are the agricultural holdings with a utilized land area less than 10 hectares. We also focused on the agroholdings with land bank above 10,000 hectares to show the increase in the proportion of large-scale land users in the Ukrainian agriculture.

To understand how agricultural land grabbing manifest itself in Ukraine, we analyzed and summarized not only official statistics, but also data provided by the independent Ukrainian organizations of agricultural producers like the Ukrainian Agribusiness Club.

Since international capital plays important role in land concentration and land grabbing worldwide, we analysed land transactions concluded by foreign investors in Ukrainian agriculture and defined land areas under their control.

We paid special attention to analyse the land raiding as an extreme form of land grabbing in the realities of the Ukrainian agricultural sector. We used a monographic method to study cases of land raiding and provided evidence of raider seizures in Ukrainian agriculture.

To understand whether any characteristics or features of land concentration and grabbing are special for Ukraine and whether similar processes are also common in countries with many land regulations to protect rural areas and small family farms, we provided a brief overview of the situation in the EU-countries and compared it with Ukrainian situation.

Then, to provide a summarized portrait of the processes related to land concentration, grabbing and raiding in Ukraine, we summarized their drivers, consequences, and other characteristics.

To define the short-term opportunities to deal with extreme land concentration and grabbing in Ukrainian agriculture under the ineffective official agrarian policy, we checked the evidences of countering land grabbing and land raiding in Ukraine.

RESULTS

Land Concentration and Land Grabbing in Ukraine

Table 1 presents the farmland distribution in Ukraine across agricultural holdings (enterprises) depending on their size. According to the State Statistics Service of Ukraine, the number of agricultural holdings with the land bank more than 10,000 hectares has increased from 93 to 166 units between 2008 and 2017. At the same time, the number of smallest farms decreased as follows: the number of farms with land area less than 5 hectares decreases by over 47 percent (2827 units) from 5965 to 3138 units, with land area between 5.1 to 10 hectares - by over 38 percent (1619 units) from 4213 to 2594 units, with land area between 10.1 and 20 hectares - by over 23 percent (1233 units) from 5170 to 3937 units, and with land area between 20.1 and 50 hectares - by over 20 percent (2855 units) from 14118 to 11263 units (*Statistical Yearbook "Agriculture Of Ukraine For 2008"* 2009; *Statistical Yearbook "Agriculture Of Ukraine For 2017"* 2018). Land areas under the control of small farms changed accordingly.

According to the State Statistics Service of Ukraine, in 2007, 93 agricultural holdings with land bank above 10,000 hectares controlled about 1.721 million hectares or 8.1 percent of all agricultural land used by agricultural enterprises. In 2018, 166 such agricultural holdings controlled 3.643 million hectares or 18.3 percent of all agricultural land used by agricultural enterprises that are legal units.

It is necessary to mention that official statistics underestimates true land banks of the largest agricultural holdings, as the State Statistics Service of Ukraine takes into account legal units and their separate subdivisions separately. In reality, many subdivisions are non-independent, but are a part of larger vertically integrated agribusiness.

According to the calculations of the Ukrainian Agribusiness Club, the business association which representing interests of the leading companies of the Ukrainian agro-food sector, in 2018, agricultural holdings with land bank above 10,000 hectares controlled 6.25 million hectares or around one third of all agricultural land used by agricultural enterprises that are legal units (*Large Farm Management Book* 2018). Such estimates are significantly higher comparing with official statistics.

Table 1. Distribution of farmland across agricultural holdings depending on their size

Farm size class (ha)	Area of agricultural land, thsd. ha		Percentage to total area	
	2008	2017	2008	2017
Holdings (enterprises) that had agricultural land	21,228.8	19960.2	100.0	100.0
including of land, ha				
no more than 5.0	19.0	10.1	0.1	0.1
5.1–10.0	33.5	20.3	0.2	0.1
10.1–20.0	79.9	61.0	0.4	0.3
20.1–50.0	536.1	424.9	2.5	2.1
50.1–100.0	348.6	354.3	1.6	1.8
100.1–500.0	1,832.1	1,797.1	8.6	9.0
500.1–1,000.0	2,049.7	1,891.4	9.7	9.5
1,000.1–2,000.0	4,090.4	3,570.9	19.3	17.8
2,000.1–3,000.0	3,338.0	2,649.2	15.7	13.3
3,000.1–4,000.0	2,481.0	1,635.4	11.7	8.2
4,000.1–5,000.0	1,659.7	1,236.1	7.8	6.2
5,000.1–7,000.0	1,822.7	1,526.3	8.6	7.6
7,000.1–10,000.0	1,216.4	1,140.1	5.7	5.7
more than 10,000.0	1,721.7	3,643.1	8.1	18.3

Source: State Statistics Service of Ukraine.

In addition to the above-mentioned farms that are legal units, there are also 4.6 million of rural households that own or use the land. They are not registered legal units. Corporate agricultural holdings use about a half of the agricultural land of rural household through rental arrangements (*Statistical Information "Main Agricultural Characteristics of Households in Rural Area In 2018"* 2018). Thus, rural households, being the landowners, do not control a significant part of their agricultural land as agricultural producers.

Following the approach of *the* Global Network for the Right to Food and Nutrition (Global Network for the Right to Food and Nutrition 2019), we calculated the *degree of land concentration as* the percentage of land controlled by small-scale farms. According to the above-mentioned approach, "small-scale" farm, in the European context, is the agricultural holdings with a utilized land area less than 10 hectares. In Ukraine, agricultural holdings with a utilized land area less than 10 hectares controlled 0.3 percent of total agricultural area in 2008, and 0.2 percent of total agricultural area in 2017. Comparing with the share of land controlled by agricultural holdings with land area above 10,000 hectares, this rate demonstrates the unequal land distribution and the very high concentration of agricultural land.

The process of land concentration also continues in 2019. The largest agricultural holdings control hundreds of thousands of hectares of agricultural land: Kernel – around 600,000 hectares, UkrLandFarmingabout – about 570,000 hectares, Agroprosperis – above 400,000 hectares, MHP – about 360,000 hectares, Astarta – about 250,000 hectares, etc. (*Large Farm Management Book* 2018).

Raising Interest of Foreign Investors for Control over Agricultural Land in Ukraine

During the last decade, we observed the growing interest of foreign investors in large-scale land acquisition in Ukraine. Unfortunately, there is a lack of official monitoring of large-scale transactions with agricultural lands in Ukraine. One of the international agencies providing data on the topic is Land Matrix that is an independent global land monitoring initiative promoting transparency and accountability in decisions over large-scale land acquisitions in low- and middle-income countries across the world.

As of June 2018, the Land Matrix had information about 123 large-scale land deals or almost 2.6 million hectares of Ukrainian agricultural land that is controlled by foreign investors from countries as follows: Austria, Belize, Belgium, Denmark, Estonia, India, Canada, Cyprus, China, Luxembourg, the Netherlands, Germany, Saudi Arabia, Singapore, United States, France and Sweden. Investors that are affiliated with Luxembourg controlled almost 800 thousand hectares, Cyprus – 650 thousand hectares, United States – 500 thousand hectares, the Netherlands –300 thousand hectares. The largest single deal was signed with an investor) from United States (430 thousand hectares). Other largest separate deals were concluded with investors from Luxembourg (two deals to control almost 350 thousand hectares and 109 thousand hectares), the Netherlands (250 thousand hectares), Cyprus (three deals to control 184 thousand hectares, 180 thousand hectares and 150 thousand hectares), and Austria (96 thousand hectares).

As of mid-August 2019, the Land Matrix has information about 213 large-scale land deals or 4,342,035 hectares of agricultural land that is controlled by foreign investors in Ukraine ("Ukraine - LAND MATRIX" 2019). It implies that the agricultural land area that is controlled by foreign investors has almost doubled in the past year.

Land Raiding as an Extreme Form of Land Grabbing in Ukrainian Agriculture

The problem of distinguishing between land grabbing and land concentration is still debatable in an academic environment. But in Ukrainian agribusiness, land redistribution and fights for agricultural land are also followed by other clearly illegal or semi-legal processes that can be called "agricultural land raiding."

Over the last decade, a whole network of companies provisioning raider services in the land tenure system has established in Ukraine. The reason behind the demand for raider services is deemed to be the realization of "global" property redistribution during the transformation of the economic system and the imperfection of the institutional changes in Ukraine. Since the early 2000s, the agrarian business began to demonstrate persistently high returns, attracting the criminal world interest. Following 2010, demand for raider seizures of agricultural enterprises and their land has become more concerned, organized and mass phenomenon. This is because of the imperfection of legislation in force, in particular, land law, the corruptness of executive and judicial power, political instability and property redistribution between financial and industrial groups, and the import of raider practices. Subsequently, farming households and even ordinary peasants' land plots became a subject to raider attacks.

One of the most popular land raiding schemes is the introduction of illegal modifications in the State Registry to establish a new owner of property and land. Registrars and notaries are engaged in organizing land raiding. Notaries make changes to the state registries in favor of raiders referring to counterfeit and questioned documents (in particular, these may be judicial documents). These light modifications in State Registry allow raiders becoming owners of a whole agrarian enterprise or its director. Consequently, the legitimate owner finds himself involved in a lengthy legal proceeding where he is forced to prove the fact of illegal modifications in the State Registry. His harvest is being sold instead and the land is not cultivated until the conflict is resolved. This scheme is commonly used to seize large-sized enterprises.

Another raiding scheme relates to leased land, that is land plots of private individuals. Several approaches are used in this case. First, the alteration or forgery of the lease agreement use rights in the registry system involving unfair state registrars and notaries. Notaries make changes to land lease registries in favor of raiders referring to counterfeit documents. Second, the registration of double lease agreements for already rented land plots. In 2013, changes in the system of land right registration were implemented and registration functions transferred from the land authorities to the judicial authorities. Right from its "launch," all cadastral records up to 2013 had not been transferred to the Registry of Ministry of Justice. Thus, the implementation of new raiding schemes proceeded by overwriting already registered agreements with a new one where another land user is mentioned. The Law on the Registration of Real Property Rights was supplemented by a rule obliging the Cabinet of Ministers to transfer all the information about the previous land rights to the new registry. This norm lasted for six months, but had not come into force. At the next modification of the law, this rule was excluded. Third, the forgery of lease agreements, including signatures and the unauthorized reassignment of agreements with other land users without the consent of landowners.

People may be aware of the fact that someone has signed a lease agreement with them as if they are paid for the land plots. Studying these documents, in fact, shows that signatures are absolutely false.

In seeking to seize land tracts of small and medium-sized farmers renting this land from fellow villagers the raiders corrupt individual landowners forcing them to terminate lease agreements and promising, in so doing, a higher rent. If a farmer keeps operating the land, he expends a lot of money on it and, consequently, goes into bankruptcy or forcibly gives the land to raiders. The promises of new land users regarding higher lease rates, in this case, are not fulfilled.

In 2016, raiding schemes on the basis of emphyteusis contracts were put into practice to seize peasants' land. According to these contracts, giving the rights of land use to a legal entity for up to 49 years, the peasant remains a landowner. During the established period of time, the land user can command over rented land on behalf of the owner. It includes leasing, disposition, borrowing money at financial institutions using leased land as a guaranty. When the company goes bankrupt, the land will become the property of a financial institution. Peasant loses the title to the land.

Those peasants who sign such contracts are poorly informed of the implications and do not understand what they accept. This scheme of land dispossession was publicly exposed and condemned by the academic community (Heyets et al. 2016). The scheme seems to be further applied with ill-informed landowners.

Rude schemes related to physical harassment and intimidation of small farms are also being used to seize their land. Small and medium-sized holdings do not have enough resources to resist raiders. Territorial remoteness makes them even more vulnerable to raider attacks. Farmers scarcely ever are able to pay for qualified lawyers or security structures to counter raiders. Pressure towards defenseless small farmers has significantly increased recently.

As can be seen from the above, the core of land raiding is composed of the gaps in the legislation and the defects in the functioning of the judicial and law system, the major land institution of the State Geocadastre, property rights and a land lease system, etc. Land legislation of Ukraine includes around one hundred laws and seven hundred regulations in the field of land tenure and land use, which are often controversial. There hardly any farmers or lawyers who would be well versed in this issue.

Evidences of Raider Seizures in Ukrainian Agriculture

Evidence from the Village of Velyka Kisnytcya, Yampil District,
Vinnytsia Region (Mnih 2019)

In early 2017, the lessee of agricultural land enacted a transfer of its rights to another affiliated company. Initially, the lessee did not even inform the local landowners that are peasants, each of which usually owned several hectares of land. Then in order to enforce landowners to sign new long-term contracts for 35 years, the lessee started blackmailing peasants by threatening not to pay them the rent payments related to a prior period. The violent confrontation lasted over a week.

Evidence from the Orativ District, Vinnytsia Region (Mnih 2019)

In late 2017, unknown armed persons savagely attacked and wounded farmer Ivan Lubarskyi. In the previous year, armed person shoot at another farmer Oleksandr Kolotus'kyi. Both cases are not just gangland shooting, but attempts to intimidate small farmers. Both affected farmers are civil society leaders and public persons with an active civil position.

There are also other similar cases, which are not even duly registered, when armed raiders put pressure on the heads of small farms in order get access to their land owned by these farmers or leased from other local peasants.

Evidences from the Public Union "Ukrainian Agri Council"
(Marchuk 2019)

The Public Union "Ukrainian Agrarian Council" counted about 600 incidents of the illicit takeover of business in the agricultural sector. The incidents were more common over the field season, from spring to fall. There are three most common raiding practices as follows:

1. Forcible harvesting crops

Raiders come to the harvest field and secure the perimeter. They bring their own harvesters and agricultural machines. Over one night, for instance, they can harvest crops over an area of 100-200 hectares. This approach is used during harvest season.

2. Forcible sowing crops

During sowing time, raiders sow the field of other farmers or lessees whose rights are violated. Then sides start a dispute at the court. As a result, the legal land user is not allowed to sow that field.

3. Seizing the farming business by fraud

Raiders seize the whole farming business, not only crops or land. It starts with falsification of documents, for example, falsification of shareholders' resolution that they agree to sell or hand over their farming business. A corrupted state of municipal registrar enters the data to the State Register for Real Estate Property Rights. Then raiders apply to the court that recognizes ownership of new owners. Some raiders, even without applying to the court, take the extract from the register and go directly to the farm. Then they take control over the farm by force and secure the perimeter of the farm.

Evidences from the Public Organization "Self-Devence
of Entrepreneurs" (Miroshnychenko 2017)

In 2017, the Public Organization "Self-Defence of Entrepreneus" promulgated information about over one and half dozen of raider seizures aimed at agricultural business. There are also evidences of raider seizures affected several dozen agricultural companies in five Ukrainian regions. As a result, nearly every above-mentioned agricultural business sustained losses in the amount of millions of dollars.

Ukraine in the International Context

Similar processes of land concentration and land grabbing, including raising foreign investments in agricultural land, can be also seen not only in the countries of the global South but also in the EU-countries.

Farmland concentration in the EU is being appeared out of agricultural businesses decrease and increase in the average land size per farm. It has an objective character and occurred at a moderate pace during the second half of the last century. Whereas in 2013, in the 27-member EU, only 3.1% of farms controlled 52.2% of farmland in Europe, and whereas, by contrast, in 2013, 76.2% of farms had the use of only 11.2% of the agricultural land (Noichl 2017).

The processes of land concentration over the last decades have been significantly accelerated both in the "old" Member States and those who joined the EU in the last wave of the enlargement. This gave grounds many researchers to conclude the change in driving forces and nature of land use concentration, recognize the strengthening role of new factors that are not directly related to agricultural production and acknowledge the growth of agricultural land grabbing in the EU.

The wide range of land grabbing processes was studied upon requests of relevant European institutions, for example, European Parliament (Cotula 2014).

Non-governmental organizations, movements, and projects played a significant role in drawing attention to the problem of land concentration and grabbing in the EU Member States and some others. Thus, under the aegis and with the participation of national members of the Friends of the Earth Europe the study of the role and impact of European financial institutions on land grabbing processes was conducted (Pentzlin et al. 2012), the impact of resource overconsumption in European economies on increased demand for land and land grabbing outside the EU was investigated (SERI and Friends of the Earth Europe 2013); upon an initiative of European Coordination Via Campesina and Hands off the Land Network the studies in this field at the level of individual European countries and regions were coordinated (Borras Jr., Franco and van der Ploeg 2013). In addition to international or pan-European organizations, the research, education, and monitoring work concerning land grabbing is being conducted by national civil society organizations, such as the Eco-Ruralis small farmers association (Baker-Smith and Szocs-Boruss 2016) operating in Romania – the state with one of the highest level of land concentration and grabbing in the EU.

In January 2015, the EU advisory body – European Social and Economic Committee adopted the Opinion on Land Grabbing in Europe (Nurm 2015), yet again stressed the problem of concentration and grabbing of agricultural land in the European Union. In 2017, as a result of civic initiative and the support of 76 civil society organizations, a petition on the need for preservation and management of agricultural land as the public property was submitted to the European Parliament.

In 2016, the Committee Hearings in this regard discussed the monitoring need and in-depth study of the last developments in the EU agricultural land use, the CAP challenges, and related legislative initiatives (Wartena, de liens and di Pierro 2019).

The findings of the study concerning the extension of agricultural land grabbing processes over the EU accomplished in 2015 upon the initiative of the Committee on Agriculture and Rural Development of the European Parliament confirmed that such processes occurring in the EU States promote a number of significant challenges and societal threats and require an immediate response (Kay, Peuch and Franco 2015). In April 2017, the European Parliament adopted the Report on agricultural land concentration in the EU emphasizing the need to provide farmers with access to land (Noichl 2017).

Family farms have traditionally been and remain the major organizational forms of farming in the agriculture, the basis of the agrarian system in the majority of the EU countries with agricultural land use dominated by the relatively small-sized farms. Land use area of the majority of European agricultural businesses does not exceed 100 hectares, which, to a great extent, is a result of the long-term implementation of the agricultural policy aimed at family farms.

According to the European Economic and Social Committee, one percent of agricultural businesses control 20% of agricultural land in the European Union and three percent control 50%. Conversely, 80% of agricultural businesses control only 14.5% of agricultural land (Nurm 2015). Another study of the European Parliament's Committee on Agriculture and

Rural Development argues that the inequality of agricultural land distribution in the EU (Gini coefficient – 0.82) is comparable to countries such as Brazil, Colombia, and the Philippines (Kay, Peuch and Franco 2015).

The transition of control over agricultural lands to external investors, corporations, financial institutions in many cases is not statistically accounted whereas farmland is de jure recognized as land owned and used by local farmers.

According to data of the European Economic and Social Committee, 0-30% of agricultural land in Romania is controlled by investors from the EU, and up to 10% - by investors from third countries; in Hungary one million hectares of land was acquired in secret deals using capital primarily from the EU Member States; in Poland, despite the fact that tough legal restrictions on the acquisition of agricultural land by foreigners were in force until May 2016, about 200 000 hectares of agricultural land have been acquired by foreign investors, mainly from EU countries (with the mediation of persons with Polish citizenship who concluded the land deals at the agricultural land sale and purchase or lease markets on behalf of external investors) (Nurm 2015).

One of the sources about the large-scale investments for taking control of agricultural land is the Land Matrix database, created by the International Land Coalition. As of October 2017, the Land Matrix database, managed by the International Land Coalition contained data on large-scale land transactions in the EU with a total land area of more than 700 thousand hectares. All reported cases concerned the territories of states that joined the EU during the last wave of enlargement – Romania, Bulgaria, Lituania. However, similar processes also take place in other European countries.

Thus, Ukraine has many similarities with some Eastern European countries like Romania, Poland and some others. Often, the actors investing in land are investors non-traditional for agriculture, namely, traders, investment and pension funds, etc. In the EU-countries, it resulted in a number of negative social, economic and ecological outcomes as follows: the marginalization of family farms, barriers for new or young farmers, erasing small landowners, threats to food security and food sovereignty, increasing rural unemployment, land degradation and others. The situation in Ukraine is similar, but it is more pronounced. Furthermore, the situation in Ukraine is deteriorating at an accelerated pace. The most extreme form of land grabbing in Ukraine is land raiding.

Summarizing the Characteristics of Land Concentration and Land Grabbing in Ukraine

We identified the main drivers and characteristics of excessive land concentration, land grabbing, and land raiding in Ukrainian agriculture.

The first point is on developing image of Ukraine as a country with underutilized agricultural lands that can be better managed with additional investments. According to the information provided by the Ukrainian Independent Information Agency of News on July 14, 2015, the former Minister of Agrarian Policy and Food of Ukraine Oleksiy Pavlenko stated that the Ukrainian agricultural sector is only running at 50 percent of its capacity. He emphasized the need for partners and investments (Pavlenko 2015). This statement may be cited as an assertion about the need for cheap land. The need for massive corporate investment means also the need for foreign investment. The purpose of this corporate foreign investment is to ensure that Ukrainian agricultural sector becomes more competitive in the international market and markets of partner countries with no provision made for the fact that domestic farmers are the major investors in agriculture. That is why domestic farmers and their investment choices should be in the focus of the investment strategies in Ukrainian agriculture.

The second point is that the largest agricultural holdings manage hundreds of thousands of hectares of land. It is the evidence of the severe land concentration in Ukrainian agriculture. Such a concentration creates opportunities for foreign investors to grab land using illegal or semi-legal schemes.

The third point is related to the market of land lease rights. The failure to respect the rights of small landowners is widespread in the market that works even without informing landowners about a transaction with their land. Moreover, there is a lack of adequate institutional protection of landowners' rights. The cost of land lease rights is a significant component of the market value of agribusiness. Sometimes the cost of land lease rights can be upwards of 80 percent of the value. Unfortunately, small landowners almost do not benefit from that.

The fourth point is that Ukraine has an enabling environment for land concentration and grabbing by large domestic and international investors. Large agricultural holdings have better access to financial resources than family farms. They have also other better opportunities, allowing them even to benefit from market failures. The mere existence of large agricultural holdings is evidence of uncompetitive environment, non-transparent market, failures in providing some public goods, inadequate market infrastructure, lack of technologies, and poor governance.

The fifth point is deficits in land administration in Ukraine. The land cadastre and register of real property rights are still in the process of formation. As the central state authority responsible for the implementation of state policies on land relations and for land management, the State Service of Ukraine for Geodesy, Cartography and Cadastre has inventoried only 11 percent of lands by the year 2018. This authority initiated reforms to decentralize land relations, tackle the problems of the black land market, and eliminate corruption. In other words, corrupt agency would reform itself.

As to the governance and corruption level, the country's indicators reflect the embarrassing situation. Ukraine ranks 77th on the Rule of Law Index among 126 countries ("The World Justice Project Rule Of Law Index 2019" 2019) and 120th countries on the Corruption Perception Index among 180 countries ("Corruption Perceptions Index 2018" 2019).

To confront the problem of land grabbing, the national authorities should pay special attention to good land administration, security of land ownership, and transparent processes to provide legitimacy to the farms. Countries with weak land administration systems are the most attractive destinations for land investments (Arezki, Deininger and Selod 2011). The reason is that it is easy to get quick and cheap access to land under such circumstances.

The sixth point is that rural communities lost a part of their rights over lands, territories and natural resources since 2013 (*Law of Ukraine "On Changes To The Ukrainian Legislation On Delimitation Of State And Communal Land"* 2013). New changes to the Ukrainian legislation were that territorial communities own the lands only within the boundaries of their built-up area, which is about 12 percent of the total. This led to restricting the public participation in the management of the local land resources and limiting the ability of local communities to tackle the problem of land grabbing. Enhancement of the capabilities of the local civil society organizations, farmers' groups and associations is an important mechanism in the prevention of the excessive land concentration and grabbing. Therefore, public policy needs to devote special attention to recognizing the rights of local communities, providing the local population with information on the value of land, their rights, and the ways for the realization of the rights.

The seventh point concerns the transnational nature and universal character of approaches to land grabbing throughout the world. Despite the global dimension of land grabbing and its attendant risks, efforts of international organizations – including the Food and Agriculture Organization of the United Nations – resulted only in endorsement of the Principles for Responsible Agricultural Investment and the Voluntary Guidelines on the Responsible Governance of Tenure of Land. However, they are good intensions and recommendation. In order to confront global land grabbing, we need decisive and immediate action at both the level of international organization and national governments.

Countering Land Grabbing and Raider Seizures in Ukrainian Agriculture

In the short-term, there is a need to stimulate and support the small landowners' association movement in Ukraine. As the existing practice of land lease relations in Ukrainian agriculture demonstrates, if landowners act autonomously, they will have not power to claim their rights and defend their interests. Under their circumstances, powerful investors will easily obtain a control over the land of smallholders. Associations of landowners are the institutions capable to protect smallholders on the land market and contribute to preventing further extreme land concentration in agriculture.

In the absence of effective state policy to counter raider seizures, the public, NGOs started a number of initiatives in this field. The above-mentioned Public Organization "Self-Devence of Entrepreneus" is a good example in this regard. It focuses its activities on consolidating entrepreneurs to eliminate the practices used for raider seizures in Ukraine.

Ironically, even associations included large-scale agricultural holdings that greatly contributed to land concentration in Ukraine (the Public Union "Ukrainian Agrarian Council" is the example) are also active in this field, because their members also suffer from the extreme forms of land grabbing and raiding. The Public Union "Ukrainian Agrarian Council" was established in 2015. They declare their mission is to consolidate the efforts of agricultural producers to protect them from illegal land raiding, provide legal support and protection to its members, and ensures that the general public is kept informed of the facts of illegal turnovers in agriculture.

The civil society organizations play a significant role in formulating policy proposals resulted in specific activities of the Ukrainian authorities. In August 2017, Ukrainian government supported the establishment of headquarters for combating raider seizures of land and crops. Such an initiative was made possible by engaging with a large number of people's deputies and the Public Organization "Self-Devence of Entrepreneus" (Miroshnychenko 2017).

As the Minister of Justice of Ukraine acknowledged at the meeting to establish a headquarters for combating raider seizures of land and crops, to confront the raider seizures in agriculture, the Ukrainian government instructed the regional administrations, Ministry of Justice, Ministry of Internal Affairs of Ukraine, and State Service of Ukraine for Geodesy, Cartography and Cadastre to engage with agrarian associations in order to establish the headquarters for combating the raider seizures in agriculture. He also invited representatives of other authorities and non-governmental organizations to participate in the Anti-Raider Working Group, which is established and composed of the representatives of the Ministry of Justice, Ministry of Internal Affairs, and Ministry of Agrarian Policy and Food. Unfortunately, the above-mentioned Working Group has not proven very successful.

Deputy Minister of Justice Iryna Sadovska believes that the work of anti-raiding headquarters is effective and reduces the number of attacks on business: 4794 complaints concerning activity of raiders were received in 2017, in 2018 – 3718, and by mid-April 2019 – 564. The signing of several lease agreements regarding the same land plot is the most popular abusive practice. Fake court decisions, unlawful cancellations of registration, blocking of agricultural equipment from entering the fields, harvesting other people's crop, illegal alteration of cadastral data take place today (Sadovska 2019).

Public sector representatives and experts believe that anti-raider headquarters' activity cannot cardinally preclude land grabbing in as much as the de-facto loyal to the raiders legislative field has not changed and headquarters include people covering illegal actions. At the same time, the illegal actions "at the bottom" do not find an adequate response at the level of the authorities, especially in the Committee on Agrarian Policy and Land Relations of the Verkhovna Rada, the Ministry of Justice of Ukraine.

In the fall of 2018, civil society, the academic community, and farmers' association with the support of the People's Deputies of Ukraine - members of the Committee on Judicial Policy and Justice submitted the new draft law No. 6236 / П to Verkhovna Rada of Ukraine on the Draft Resolution for Further Work on the Draft Law amending Article 41 of the Constitution of Ukraine on the Realization of Ukrainian Citizens Rights to Land, the Preservation of Agricultural Land Ownership, and the Sustainable Rural Development on the basis of Farming Households and forwarded it to the Constitutional Court for obtaining its opinion. It was planned the draft law to amend Article 41 of the Constitution with a new part: "Farming household is the basis of the agrarian system of Ukraine." At the same time, the title of the draft law states that the changes are related to the realization of Ukrainian citizens' rights to land, the preservation of ownership of agricultural land, and the support of sustainable rural development. *In January 2019,* a draft resolution on further work on this draft law (Knyazevych 2019) was registered in parliament, but People's Deputies rejected the resolution in May.

The Ministry of Justice of Ukraine advises agricultural producers to fend for their property without relying on the state. The First Deputy Minister of Justice reported in April 2019, "As of today, every owner can protect their business using existing tools. ... I urge everyone not to wait for new laws and not to rely solely on the state" (Sukmanova 2019). The Ministry of Justice published traces of a possible raider attack that should alert the business owner: incoming of envelopes with blank letters or advertising; increasing the number of different checks as an effort to obtain the maximum number of documents; dissemination of negative information about the agricultural enterprise in the media; lawsuits on various grounds; offers to sell the business, even if the owner does not plan to sell it; the purchase of the enterprise debts and frequent request for information on the registry of property rights and leases of agricultural enterprises, etc. Olga Matviyiv, director of the Business-Warta, responded to these tips as follows, "People work thinking that the State protects them. However, it turns out that they have to constantly monitor the legislation and check the State Registry whether they have already been robbed. Business more than ever talks openly about problems in government control..." (Petrenko 2019).

Bureaucratic, corrupt and inefficient management of land resources performed by the major land department of the country - the State Service of Ukraine for Geodesy, Cartography and Cadastre (Geocadastre) facilitate land grabbing. During the period of continuous reorganization, this agency has not changed much in the last two decades and continues to concentrate its power uncontrollably.

Maintenance of the State Land Cadastre is an extremely important state function assigned to the State Geocadastre service. According to experts of the NGO "Land Union of Ukraine," huge funds are constantly being spent on its financing and administration, but the land cadastre remains undeveloped. The State Service of Ukraine for Geodesy, Cartography, and Cadastre - the fifth-largest by staff civil service in Ukraine. Taxpayers spent 1.5 billion hrivnas in 2018 for its maintenance. However, according to the agency, the state land cadastre contained information about only 75% of the land at the beginning of 2019 (the process of filling the cadastre takes almost 20 years). Private lands are the best registered, the worst - state and communal lands.

Most users of cadastral information do not have normal access to it. A publicly available cadastral map is still a simple picture. The State Geocadastre does not fulfill the direct requirements of the law on the publicity of cadastral information and strongly refuses to share it. Furthermore, this agency was granted the rights to manage state lands in 2013 (*Law of Ukraine "On Changes to the Ukrainian Legislation on Delimitation of State and Communal Land"* 2013), thereby created an uncontrolled "closed cycle" of state land redistribution. "According to the legislation, the State Geocadastre manages the land itself, approves the land management documentation, maintains the land cadastre, conducts land valuation, manages a large number of commercial state-owned enterprises and, even more, formally monitors itself as the government has entrusted it with government control over land use and protection" (Martyn 2019). According to experts, "Land relations should be subject to systematic simplification and deregulation. The surpluses of the power and responsibilities of the land office should be organically distributed among other central executive authorities and local self-government bodies. The lion's share of land management powers should be delegated to local communities. The cadastre should be decentralized in conjunction with the registration of rights. The responsibility for geodesy and cartography should be transferred to the military for a long time. The registrars should work in local governments" (Martyn 2019).

Land reform can significantly improve the situation of land grabbing and forward these processes into a civilized bedrock.

CONCLUSION

Land concentrations slowly started in 1990s in Ukraine, but in the next decade, it continued at an alarming rate and began to assume a character of land grabbing. Now, there are evidences of the continuing land concentration, land grabbing and its extreme form, namely, land raiding. There are tendencies to monopolizing control over agricultural lands by large agrarian holdings in many regions. It decreases the viability of both agricultural sector and rural areas. Exclusion of agricultural land from local and rural development, illegal land market transactions, and failing to meet the principles of responsible investments are the main features of land use by large corporate holders.

The defining characteristic of the modern Ukrainian land use in agriculture is an imbalance between the interests of large agricultural holdings and the public interests. The system of agricultural land use is not aimed at realizing sustainable development, smart specialization of rural areas, development of local markets, ensuring biodiversity, developing the potential of local economies and their diversification.

In Ukraine, there are indications of land grabbing as follows:

- Large-scale land acquisitions and long-term-leases;
- Industrial agriculture with devastating effects on ecosystems;
- The exceptionally large scale of displacement of the rural poor and low-income groups from farming and land relations without any compensations;
- Participation of foreign investors in large-scale purchases of rights on land lease, their control over land;
- Participation of national elites in large-scale land lease and land purchases using semi-legal and non-transparent practices;
- Large-scale land use for export-oriented monoculture production;
- Reducing the opportunities for food self-sufficiency;
- Land concentration is colonial in nature (producing food for neither local consumption nor benefit of local communities);
- Usage of illegal or semi-legal practices for large-scale land transactions;

- Restrictions on human rights and limited livelihood opportunities in rural areas by seizing control over decisions on land use, periods of land use, and security of land tenure;
- Public control over large-scale agricultural land use is restricted;
- Access to agricultural land for most vulnerable and marginalized groups is restricted;
- There are lots of evidences of land raiding in agriculture.

The main drivers of land concentration and grabbing in Ukraine are the following:

- Decision-making processes are non-transparent;
- Corruption schemes and practices are common;
- High level of poverty is common in rural areas;
- The level of public participation and influence on political and societal processes is low;
- Land tenure and land use systems are not safe;
- As to state agricultural lands, the governance is weak and the system for effective control is absent;
- The infrastructure is poor, producer and financial service sectors are weak and force the rural population to abandon the use of agricultural lands.

Land concentration and land grabbing in Ukraine resulted in a number of negative social, economic and ecological outcomes. The main consequences of land concentration and grabbing in Ukraine are the following:

- A small number of large-scale agricultural holdings has concentrated control over a significant part of agricultural lands;
- New forms of land use control, which are colonial in nature, became common in Ukraine;

- The Ukrainian agricultural sector has been reformatted and corporatized;
- There are many negative changes in social and environmental attitudes related to agriculture;
- Large-scale investments in agricultural land lead to the dispossession of small farmers, disappearance of small farms, violations of the rights of peasant and other people living in rural areas;
- Ukrainian family farming is becoming marginalized. There are many barriers for new and young farmers;
- There are new threats to food security and food sovereignty;
- Rural unemployment continues to increase;
- Land and soil degradation became a fact.

There is a need in this area for stronger agricultural policy actions for both reducing the negative effect of large-scale corporate land use and supporting family farming that is characterized by more sustainable, socially and ecologically responsible land use. In the short-term, there is also a need to stimulate and support the small landowners' association movement for protecting smallholders on the land market and preventing further extreme land concentration in agriculture.

Participants

Olena Borodina
Viltor Yarovyi.

REFERENCES

Arezki, Rabah, Klaus Deininger, and Harris Selod. 2011. *What Drives The Global Land Rush?* International Monetary Fund.

Baker-Smith, Katelyn, and Miklos-Attila Szocs-Boruss. 2016. *What is Land Grabbing? A critical review of existing definitions.* Eco Ruralis. https://drive.google.com/file/d/0B_x-9XeYoYkWSDh3dGk3SVh2c Dg/view.

Borras Jr., Saturnino, Jennifer Franco, and Jan Douwe van der Ploeg. 2013. *Land Concentration, Land Grabbing and People's Struggles in Europe.* Amsterdam: Transnational Institute. https://www.tni.org/files/ download/land_in_europe-jun2013.pdf.

"*Corruption Perceptions Index 2018.*" 2019. https://ti-ukraine.org/en/ research/corruption-perceptions-index-2018/.

Cotula, Lorenzo. 2014. *Addressing The Human Rights Impacts Of 'Land Grabbing.'* Brussels: European Parliament. http://www.europarl. europa.eu/thinktank/en/document.html?reference=EXPO_STU(2014)5 34984.

Dell'Angelo, Jampel, Paolo D'Odorico, Maria Cristina Rulli, and Philippe Marchand. 2017. "The Tragedy of the Grabbed Commons: Coercion and Dispossession in the Global Land Rush." *World Development* 92: 1-12.

European Coordination via Campesina. 2016. "How Do We Define Land Grabbing?" Brussels: European Coordination via Campesina. https://www.eurovia.org/how-do-we-define-land-grabbing/.

Global Network for the Right to Food and Nutrition. 2019. "Degree of Land Concentration." *Righttofoodandnutrition.Org.* Accessed August 19. https://www.righttofoodandnutrition.org/degree-land-concentration -15.

Heyets, Valeriy, Olena Borodina, Lubov Moldavan, Yuriy Lupenko, Volodymyr Yurchyshyn, Vasyl Andriychuk, and Tamara Ostashko et al. 2016. "Who Benefits From The "Improvement Of Lease Relations" Based On Emphysesis." *Dzerkalo Tyzhnya*, 2016. https://dt.ua/ macrolevel/komu-vigidne-udoskonalennya-orendnih-vidnosin-na- osnovi-emfitevzisu-_.html.

Kay, Sylvia, Jonathan Peuch, and Jennifer Franco. 2015. "Exnent of Farmland Grabbing in the EU." Brussels: European Parliament. *Policy Department B: Structural and Cohesion Policies.* http://www.europarl.

europa.eu/RegData/etudes/STUD/2015/540369/IPOL_STU(2015)5403 69_EN.pdf.

Knyazevych, Ruslan. 2019. "Draft Decision on Further Work on a Draft Low on Changes to Article 41 of the Ukrainian Constitution." *Werkhovna Rada of Ukraine.* http://w1.c1.rada.gov.ua/pls/zweb2/ webproc4_1?pf3511=65374.

Large Farm Management Book. 2018. Ukrainian Agribusiness Club.

Law of Ukraine "On Changes to the Ukrainian Legislation on Delimitation of State and Communal Land." 2013. https://zakon.rada.gov.ua/laws/ show/5245-17: Verkhovna Rada of Ukraine.

Law of Ukraine "On Disposal of Land Share." 2001. https://zakon. rada.gov.ua/laws/show/2242-14: Verkhovna Rada of Ukraine.

Martyn, Andriy. 2019. "State Geocadastre tricks - what's wrong with the 'land' department?" *Agropolit.com.* https://agropolit.com/ INTERVIEW/535-ANDRIY-MARTIN-FOKUSI-DERJGEOKADA STRU--SCHO-NE-TAK-U-NAVKOLOZEMELNOMU-VIDO MSTVI.

Mnih, Antonina. 2019. "Agricultural Raiding: Would The Incident Rooms Protect Agricultural Producers?" *Ukrinform.Ua.* https://www. ukrinform.ua/rubric-economy/2301605-agrarne-rejderstvo-ci-zahistat-silgospvirobnikiv-operativni-stabi.html.

Marchuk, Denys. 2019. "The Top 3 Practices of Raider Seizure in Agricultural Complex." *Agravery Agrarian News Agency.* http://www. agravery.com/uk/posts/show/top-3-shemi-rejderskih-zahoplen-v-apk.

Miroshnychenko, Ivan. 2017. "Who's In Charge Of Raider Seizure in the Ukrainian Agriculture?" *Economichna Pravda.* https://www.epravda. com.ua/columns/2017/08/11/627972.

"New Ukrainian President Wants to Allow The Sell of Agricultural Land by the End of 2019." 2019. *East-Fruit.Com.* https://east-fruit. com/article/novyy-prezident-ukrainy-khochet-razreshit-prodazhu-selkhozzemli-do-kontsa-2019-goda.

Noichl, Maria. 2017. "Report On The State Of Play Of Farmland Concentration In EU: How To Facilitate The Access To Land For

Farmers." Brussels: European Parliament. http://www.europarl.europa. eu/doceo/document/A-8-2017-0119_EN.html.

Nurm, Kaul. 2015. "Land Grabbing – A Warning for Europe and a Threat to Family Farming." Brussels: European Economic and Social Committee. https://www.eesc.europa.eu/en/our-work/opinions-information-reports/opinions/land-grabbing-europefamily-farming.

Panasenko, R.A. 2010. "On Criminalization and Interpretation of "Raiding" Concept In Ukraine." *Visnyk Luhanskoho derzhavnoho universytetu vnutrishnikh sprav imeni E. O. Didorenka*, no. 5: 21-29.

Pavlenko, Oleksiy. 2015. "Agricultural sector is only running at 50 percent of its capacity – the Minister." *UNIAN*, July 14, 2015. Accessed August 7, 2019. https://www.unian.ua/economics/agro/1100590-agrarniy-sektor-ukrajini-funktsionue-lishe-na-50-potujnosti-ministr.html.

Pentzlin, Daniel, Rachel Tansey, Alexander König, Antonio Tricarico, Markus Henn, Hannah Griffiths, Christine Pohl, Julia Huscher, Kenneth Haar, Andrea Baranes, Amy Horton and Robbie Blake. 2012. *Farming Money: How European Banks And Private Finance Profit From Food Speculation and Land Grabs*. Brussels: Friends of the Europe. http://www.foeeurope.org/sites/default/files/publications/ farming_money_foee_jan2012.pdf.

Petrenko, Pavlo. 2019. "Opinion: In 2019, the risk of raids is still high." *Agravery Agrarian News Agency*. http://www.agravery.com/uk/posts/ show/dumka-u-2019-zagroza-rejderskih-atak-na-biznes-zalisaetsa-visokou.

Sadovska, Iryna. 2019. "Duplication of lease agreements is the most popular practice of agricultural raiding in Ukraine." *Agravery Agrarian News Agency*. http://www.agravery.com/uk/posts/show/podvoenna-dogovoriv-orendi-najpopularnisij-vid-agrorejderstva.

SERI and Friends of the Earth Europe. 2013. *"Hidden Impacts: How Europe's Resource Overconsumption Promotes Global Land Conflicts."* Vienna: GLOBAL 2000. https://www.foeeurope.org/sites/default/files/ publications/foee_report_-_hidden_impacts_-_070313.pdf.

Shapiro, V. S., and T. V. Bulakh. 2015. "Raiding As Economic and Legal Phenomenon: General Characteristics." *Forum Prava*, no. 2: 181-186. http://nbuv.gov.ua/UJRN/FP_index.htm_2015_2_32.

Statistical Information "Main Agricultural Characteristics of Households In Rural Area In 2018." 2018. State Statistics Service of Ukraine.

Statistical Yearbook "Agriculture Of Ukraine For 2008." 2009. Kyiv: State statistics committee of Ukraine.

Statistical Yearbook "Agriculture Of Ukraine For 2017." 2018. Kyiv: State statistics committee of Ukraine.

Sukmanova, Olena. 2019. "Opinion: Every business owner has to take measures to secure business from raids." *Agravery Agrarian News Agency.* http://www.agravery.com/uk/posts/show/dumka-kozen-vlasnik-biznesu-povinen-vziti-zahodiv-dla-ubezpecenna-kompanii-vid-rejderskih-atak.

"The World Justice Project Rule of Law Index 2019." 2019. Washington, D.C.: World Justice Project. https://worldjusticeproject.org/sites/default/files/documents/WJP_RuleofLawIndex_2019_Website_reduced.pdf.

"Ukraine - LAND MATRIX." 2019. *Landmatrix.Org.* https://landmatrix.org/country/ukraine/.

Wartena, Sjoerd, Terre de liens, and Marta di Pierro. 2019. "Preserving And Managing Land As Our Common Wealth." Hearing of the Petition 0187/2015 before the Committee on Petitions of the European Parliament. Brussels. https://www.eurovia.org/wp-content/uploads/2016/07/Petition_Hearing_Statement_Wartenaal_f2.pdf.

In: Land Use Changes
Editor: Vinícius Santos Alves

ISBN: 978-1-53617-032-0
© 2020 Nova Science Publishers, Inc.

Chapter 3

SOIL FUNCTIONS AND LAND MANAGEMENT: POTENTIAL SYNERGIES AND TRADEOFFS AT THE TROPICAL AGRICULTURAL FRONTIER (BRAZIL)

Maria Victoria R. Ballester[1,], Andrea S. Garcia[1], Rodnei Rizzo[1], Vivian M. F. Vilela[1], Érica S. Nakai[1], Mayra de F. Preto[1], Laura P. Casarin[1] and Daiana M. Tourne[2]*

[1]Center of Nuclear Energy in Agriculture, University of São Paulo, Piracicaba, SP, Brazil
[2]Center for Environmental Studies and Research (NEPAM), University of Campinas, Campinas, SP, Brazil

---

[*] Corresponding Author's Email: vicky@cena.usp.br.

ABSTRACT

We developed a framework which addresses sustainable land management through ecosystem services provided by healthy soils. Our conceptual tool is designed to support policy makers to manage five selected soil functions and demands: productivity, water availability, nutrients, carbon sequestration, biodiversity. We applied it to Amazon's Agricultural frontier - the Upper Xingu River Basin, state of Mato Grosso, Brazil – one of the most rapidly changing regions in the basin. In this study we implemented an interdisciplinary, GIS-based, multi-model approach to understand how land use change due to agricultural expansion and intensification is affecting soil functions at the Amazon's Arc of Deforestation. Encompassing two large Brazilian Biomes, the tropical rain forest and the Cerrado (Brazilian Savannas), the study area of ~170.000km^2 has undergone extensive changes in land use and land cover since the late 1970s. In only 40 years, the basin has already lost 30% of its natural vegetation, which was replaced mainly by pastures for cattle ranching. Since the early 2000s, a new cycle has started and currently the Upper Xingu River Basin is undergoing an increased agriculture intensification process (e.g., double cropping) to produce corn and soy bean for the international market. According to stakeholders, the main limiting factor for agriculture is infrastructure and absence of government. The lack of support affects markets and international trading with high costs for stocking and distribution of soy bean products. As proxies for soil indicator for supply we used a 2015 land use map derived from remote sensing data. Demand was expressed as annual productivity (kg/ha/y) from census data spatialized by land use type. Evapotranspiration was used as the proxy for water availability and water yield for demands. For a biodiversity indicator we used Indigenous Land and Conservation Units as suppliers and Legal Reserve and Permanent Protection Areas as demand (legal instruments). Nutrient Cycling supply was mapped based on a base saturation of 0-30 cm of almost 400 soil profiles and demand was mapped using average fertilizer application (NPK) spatialized by land use. Carbon storage and sequestration were derived from field measurements spatialized by land use and demands were generated using field carbon sequestration measurements and policy (Zero illegal deforestation target). Our results show that this approach can be applied to a range of landscapes and is a useful tool for decision making and policy implementation and support.

Keywords: land managenment, land use, ecosystem services, soil functions

INTRODUCTION

In recent decades, multiple, complex and interactive human activities have changed the shape of the Earth System fast and profoundly. Our imprint spreads on the global environment, affecting ecosystems and societies at all scales, raising concerns on how these changes may affect ecosystems and human well-being. In the 21's century, humanity faces many and multifaceted challenges to ensure water, food and energy security for a growing population in a climate changing world. Meeting current and future demand for goods such as food, water, fibre, energy and other resources are the main drivers of land cover and land use change (Foley et al. 2011). In addition, conversion of natural land cover into anthropogenic landscapes increases economic income and employment opportunities.

Land use changes encompass a series of different processes, ranging from initial natural vegetation clearance to land management modification. Currently, as commodities trade has become a global phenomenon, country's demands for feedstocks, associated with national and international policies to provide them, have progress from a local/regional to global drivers of land use change. Therefore, the expansion of agricultural land and the intensification of agricultural production are among the most profound human alterations of the global environment. Projected growth for 2050 of cereals demands for both food and animal feed uses are almost 1 billion tones, which will be associated with a considerable expansion of international trade and production intensification (FAO, 2009). Earth's tropical regions are now the frontier for this global trend. Most of the expansion and intensification of the world's agricultural production over the next several decades are projected to take place in the tropics. This intensification can involve both changes to cropping systems and the integration of regional animal and crop production.

Forest cleaning can lead to impacts on ecosystems goods and services, such as decrease in evapotranspiration (Coe et al. 2011) and changes in regional rainfall and river discharge patterns (Rizzo et al. subm.), increases in soil temperature, erosion and modifications in nutrient availability (Ballester et al. 2013; Deegan et al. 2011) and alteration in river transport of sediments, organic matter and associated nutrients and biodiversity (Deegan et al. 2013). Since in these areas usually food production relies on climatic stability and soil functions, when combined with climate change, current threats to economic activities and human wellbeing are likely to be even more intense (Pielke, 2005; Alves, 2007; Macedo et al. 2012).

Soil functions refer to soil based ecosystem services, an overarching concept referring to one out of five (Schulte et al. 2014) elemental aspect of the soil system that contributes to the generation of goods and services (Landmak, 2019). All soils perform these functions simultaneously, but both the extent and the relative composition of this functionality depends upon pedological, physical, chemical and biological soil properties (Shulte et al. 2014; O'Sulivan et al. 2018). The contemporary principal soil functions pertaining to agricultural land use and cattle ranging are primary productivity, water purification and regulation, carbon sequestration and other aspects of climate regulation, provision of a habitat for functional and intrinsic biodiversity and nutrient cycling and provision.

In the Brazilian Amazon arch of deforestation, increase in food production have been associated with a fast and unprecedented spread of deforestation and land use change (Batistela and Moran, 2007; Norton et al. 2006; Garcia et al. 2017) in the last four decades. While in 1970's and 1980's the expansion of the agricultural frontier was mostly drive by governmental policies that promoted settlement process, financing and infrastructure development (Coy and Kohlhepp, 2005; Mello et al. 2006), since the 1990's, teleconnections have become more important (Rausch and Gibbs, 2016). Foremost production expansion and intensification in some regions such as in Mato Grosso State are under a dynamic process in agriculture today, with a rapid technological modernization that impacts rural productive structures, with strong reflexes on land tenure and soil functions. Pressing problems that are now emerging concern different challenges related to adaptation to climate change and the landscape ability to provide ecosystem services. One clear example is the decrease in regional precipitation and river discharge found at the Upper Xingu River basin (Rizzo et al. sub).

Concurrently, landscape dynamic studies have mostly concentrated on the tropical rain forest, while forest-savannas ecotones and the Cerrado biome, one of the largest and richest savannas in the world that is under the highest rates of land use conversion in Brazil (Garcia & Ballester, 2016), have receive less attention. Therefore, there is a lack of information on how the landscape dynamic is been shaped by the couple effect of bio-phisycal characteristics (e.g., climate, topography, soils, etc) and anthropogenic factors (Garcia et al. 2017; Sawakushi et al. 2013), such as cultural heritage and environmental perception, institutional arrangements, governmental and private policies for financing technological incorporation and infrastructure development, settlement process, land tenure and international trade.

Understanding land use change drivers associated with past and current land management and historical process of colonization are crucial to provide adequate scientific information for soil function sustainable management. Here we present the results of an integrated spatially explicit temporal analysis of synergies and trade off of land management and soil functions by applying the Functional Land management Approach in a GIS modeling framework to a landscape located at the agricultural frontier in Brazil.

METHODS

Study Area

The Upper Xingu River Basin is located at Brazil's central west region, at the Amazon deforestation arch (Figure 1). Draining an area of almost 170.000 km^2 and encompassing 0.03% of the world's and 1,6% of Brazil's agricultural lands, this basin is highly socio-environmentally diverse, playing an important role in national and international commodities markets. About 2% of the world's and 9% of Brazil's soybean crop is cultivated in the region, while 0.4% of the world's and 3.0% of Brazil's cattle herd are grown in the area (FAO 2019; IBGE 2017a). The landscape is an intricate mosaic of natural and human introduced patches, located in the transition area between the Amazon tropical rain forest and the Brazilian savannah named Cerrado (Figure 1). Average temperature ranges from 18 to 36 oC, while total annual rainfall ranges from 1400 to 2500 mm. The original vegetation of the basin was composed of 65% tropical rain forest, 25% Cerrado (Brazilian savannah), 2% Forest to Cerrado transition and Riparian Forest, respectively, and 6% pioneer vegetation (IBGE 1988).

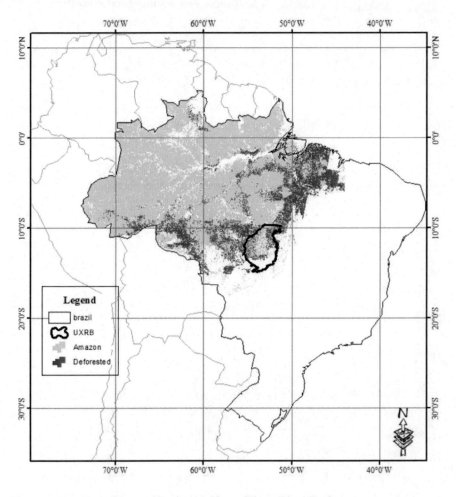

Figure 1. Location of the study area, the Upper Xingu River Basin, Mato Grosso, Brazil.

GIS Analysis and Scenarios

Using Geographical Information System ArcGis 10.6 (ESRI 2017) software, a GIS-Based multi-model was developed to support governance and policy making to manage 5 soil function supplies and demands. Our GIS template consisted of a geodatabase compiled as a digital collection of geographic datasets of biophysical and anthropogenic landscape characteristics and drivers that we used to implement four different scenarios of soil functions and land management supply and demands (Table 1).

Selected biophysical factors included a 1:5.000.000 soil type map (Santos et al. 2011) and soil profiles (EMBRAPA 2014). To obtain a continuous soil texture surface, data from 0 – 30 cm depth profiles was interpolated by the inverse square distance (De Castro Victoria et al. 2007) for percent of clay, sand and silt content, base saturation and cation exchange capacity. Spatially integrated average and standard deviation values were derived from all digital numeric models, using as zone layers either soil units, basin or property level.

A land tenure dataset was derived from an assemblage of three digital layers: a) indigenous reserves and conservation unit limits, acquired from the Brazilian Ministry for the Environment (MMA 2015); b) farms boundaries, obtained from the environmental rural registry (CAR) of Mato Grosso State (SEMA-MT 2016), and c) rural settlement limits from the National Institute of Agrarian Reform (INCRA 2016). Each data set was first georeferenced to a high-resolution satellite image mosaic of the upper Xingu Basin, derived from 380 Rapid Eye 5-meter resolution scenes retrieved from MMA (http://geocatalogo.mma.gov.br).

Next, a series of semi-automatic and manual corrections of double boundaries, overshot and undershot limits were necessary. The analysis of final combined land tenure map was produced comparing five major groups of stakeholders: indigenous people, small rural producers (<100 ha), small farmers (>100 and <400 ha), medium farmers (>400 <1500 ha) and large farmers (>1500 ha).

Table 1. Soil function, supply and demand data sources

Soil function	Supply	Demand	Data Source
Primary Productivity	2015 land use map classified as Kg of protein /ha/year according to regional census average values	Brazil's governmental 2025 decadal targets for soy, corn and cattle converted into Kg of protein /ha/year degraded pastures converted to soybean/corn secondary growth converted into pasture	(Garcia et al. 2019; IBGE 2017a; Brazil 2018)
Water availability	Water yield	Vegetation, including crops and pasture, average water demand	(USGS 2019)
Nutrients demands	Natural abundance from soil profiles	Regional fertilization practices	(EMBRAPA 2014)
Carbon sequestration	Land use map and literature	Literature and management practices	(Garcia et al. 2019; Corbeels et al. 2016; Miranda et al. 2016; Raucci et al. 2015)
Biodiversity	Indigenous land Permanent Protection Areas	Law 12.651/2012	(Brazil 2012)

A time series from 1990 to 2014 of census data by county included: number of inhabitants, planted and harvested areas, productivity, number of cattle herds obtained from the Brazilian Institute of Geography and Statics (IBGE 2017a). Data on the amount of corn, meat and soybeans exported, destination markets, and the commercial value of exported commodities were acquired from the System of Analysis of Foreign Trade Information (AliceWeb2 2016). Our spatially distributed analysis for these data sets took into account boundary changes during our time frame analysis window.

Settlement process data was derived from census data acquired from the Brazilian Institute of Geography and Statistics (www.ibge.gov.br), while historical data was obtained from the literature, legislation, stakeholder interviews and analysis of public policy plans. Current land tenure maps were obtained from: the State Environmental Secretary of Mato Grosso (www.sema.gov.mt) and from the National Agrarian Reform Ministry of Brazil (www.incra.gov.br) for farms and governmental settlement projects;

the Brazilian Ministry of Environment (www.mma.gov.br) for indigenous reserve limits, road networks, urban centers, and large reservoirs. A time series (1985–2015) of land use and cover maps was produced by satellite images digital classification. Details can be found elsewhere (Garcia et al. 2019).

To understand how land management is affecting soil functions, we first mapped demand and supply of five soils functions (Table 1). For each soil function we modeled three scenarios: a) business as usual, which includes current land management practices, extensive cattle ranging in pastures areas with an average value of 1.5 heard per hectare, pasture management restricted to eventual use of fire to increase nutrient content, as well as single and double-cropping (corn-soy bean rotation), direct seeding, rain fed fields and soil pH correction by liming and NPK addition; b) conventional intensification, which includes secondary growth area use for cattle ranging, degraded pasture converted into soybean, and single cropping converted into double cropping; c) sustainable intensification of agroforestry associated with cattle ranging in degraded pastures and secondary growth areas. All scenarios were designed to meet 2025 governmental agribusiness projections (Brazil 2018). Carbon sequestration is the product of the amount of fixed carbon minus the amount of CH4 (as CO2 equivalent) emitted by cattle in each scenario.

RESULTS AND DISCUSSION

The Upper Xingu basin river network is composed of a dense (average 0.56 km.km-2) system of channels, dominated by low-order streams with a total length of almost 95,000 km and 2,147 head waters. The terrain is predominantly flat - 98% of the basin has a slope equal to or less than 12%, allowing for agricultural mechanization. Most of the soils (68.6% of the basin area) are classified as nutrient poor but physically well-structured Oxisols. On average, soil texture is 26.1% (±9.6) clay, 65.2% (±10.6) sand and 8.9% (±2.3) silt. As a consequence, head water biogeochemical characteristics are mostly very poor in ions, with electrical conductivity values (5,2 to 33,5 µS.cm^{-1}) indicating drainage of highly weathered terrains.

Mann-Kendall trend analysis and Pettit's change point detection analysis of a 40-year time series (1976 to 2015) of rainfall and river discharge showed that from 1985 to 2015, rainfall decreased by about 245 mm over the period, but there was no trend in river discharge. The number of rainy days and the intensity of rain events also decreased, but the length of the rainy season and seasonal and annual discharge did not change (Rizzo et al. 2019). The change in the rainfall trend was represented by a break in time series which was associated with the initial deforestation peak in the basin. Although we still cannot separate climate change and deforestation effects on regional water balance, there are clear signs that land use change plays a key role in this process. Forest replacing by soybean at the lower scale results in changes in water infiltration in soil profiles and river discharge. In areas draining soybean water infiltration is faster and water retention lower, resulting in four times increase in river discharge (Riskin et al. 2013).

Another important characteristic of this basin is that precipitation is heterogeneously distributed, with a spatial pattern that decreases from North West to South East and a temporal pattern of dry and wet seasons corresponding to July-October and November-March, respectively (Rizzo et al. 2019). As a consequence, evapotranspiration and water yield present the same spatial pattern, ranging from 500 to 1400 mm and 277 to 1380, respectively.

Soils in the basin are acidic and poor in nutrients, but have good physical structure. Therefore, the main practices are liming and NPK fertilizer application. This management practice is already affecting water quality as shown by (Riskin et al. 2013). In this study, river surface water that drained soy fields carried more nutrients that water draining from forested ones.

The colonization of the Upper Xingu River Basin was characterized by complex pre-Columbian (before European contact) indigenous settlements (Heckenberger et al. 2003, 2008), followed by expansion of an active pioneer frontier (Arvor et al. 2012). While other parts of Mato Grosso State started to be exploited for gold mining and agricultural production during

Brazil's colonial period (~1700), the Upper Xingu River Basin remained relatively restricted and isolated until 1884. In this year, Karl von den Steinen's expedition arrived at the basin's southern boundary and founded a territory that shared in a peculiar cultural symbiosis with four of the major Brazilian linguistic indigenous groups organized in a type of confederation (Schaden 1990), living in a highly self-organized anthropogenic landscape of late prehistoric villages and road networks across the region (Heckenberger et al. 2008). Currently, the basin is inhabited by 20.000 indigenous residents, distributed across 13 reserves encompassing about 4 million ha (or 30% of the basin) (Velasquez, C.; Alves, H. Q.; Bernasconi 2010; IBGE 2017b). This diverse indigenous population is composed of at least 06 ethnic groups: Bakairi, Wauja, Kayapó, Kisêdjê, Tapirapé e Xavante (FUNAI 2019).

The settlement process at Brazil's Amazon agricultural frontier has been associated with several public policies and economic cycles that shaped large portions of the landscape very fast. From 1945 to 1964, no land policy was in place to plan, direct or regulate territorial development. State land sales lacked precise geographical maps and definitive title deeds were indiscriminately granted for large areas. To promote the national integration and regional development of the Amazon region, the First National Development Plan (1970–1972) implemented a series of settlements for small farmers and infrastructure along two road axes: BR-364 and Transamazônica.

While slash and burn were the most widely used practices for converting natural primary forest to pasture and crops, different and intricate types of colonization and associated land management resulted in clear and distinctive landscape patters. For instance, public road building was associated with lot allocation schema, leading to a deforestation pattern known as fishbone. Settlement projects and lots were allocated in the landscape according to land availability. Moreover, biophysical characteristics, excessive distance from markets, poor infrastructure, absence of technical support and inadequate management practices for the region, were among the most important reasons for the lack of success of many settlers.

The second National Development Plan (1975–1979) shifted its target to developing agricultural and cattle ranging centers (Mello 2006). Promoted by the federal government and mainly implemented by private companies and cooperatives, colonization spread particularly in Mato Grosso and the Upper Xingu Basin. The establishment of tax exemptions and tax deductions for companies to implement rural settlements in the region were used promote internal migratory flows. Originating from the Northeast and Central-South of the country to "occupy empty spaces," this process was designed to guarantee the interconnection of the region with the rest of the country by the federal government.

As a consequence, current land tenure in the Upper Xingu River Basin is the result of a settlement policy that consisted of the sale of vast public areas of empty land to large companies at insignificant prices (Moreno, 1998). In turn, these companies developed livestock projects, by forest clearing and pasture establishment, or divided the land into lots to sale to settlers. Only in 1987, the National Institute for Agrarian Reform (INCRA) implemented the first Joint Settlement Projects (PACs), by combining the experiences and resources of the official colonizing agency (INCRA) with those of the private initiative. Road opening was also an important process which facilitates transport of people and goods, and, therefore, is a attracts more people.

Contrasting with other regions of the Amazon agricultural frontier such as Rondônia and Pará, where large colonization projects were established by the Brazilian federal government in the late 1970s and early 1980s, in the Upper Xingu Basin, private companies and cooperatives also played a key role in the settlement process. Combined with the fact that this region encompasses a large indigenous reserve, land tenure shaped the landscape, leading to different patterns and ecological processes.

Land occupation and land use change in the Upper Xingu basin is recent, spatially heterogeneous and extremely fast (Garcia et al. 2019). Although 68% of the land is still natural vegetation, 34% was converted in only 30 years (1970-2015,) mainly at the head waters of the basin. Land use change involves not only conversion from pasture to soybean fields, but also forest clearing and replacement by pasture (Leite et al. 2011; Macedo et al. 2012;

Morton et al. 2006). Aside from the extensive changes in land use that this area has undergone in the last three decades, at present, agriculture intensification has also been introduced, mainly by double cropping (soybean-corn). In the early 2000s, as market demand for soybean increased, soy cropland started to replace pastures and forests, and only five years later, double cropping became a common practice. Expansion processes were dominant from 1985 to 2005 in both cerrado and forest areas, while from 2005 to 2015, intensification became more significant (Garcia et al. 2019).

We were able to identify 23 different private companies and 01 private cooperative that operated as colonizers in the Upper Xingu River Basin. Moreover, the main processes of land acquisition in 2006 were 67% from private companies and 27% through governmental settlement programs (IBGE 2017a). Land is heterogeneously distributed among local stakeholders, with about 4 million ha (or 33% of the basin area) belonging to 13 indigenous reserves, including the Xingu Park. Another 8 million ha (47% of the area) are distributed among 1270 large farms (>1000 ha), 939 medium farms (>100 <1000 ha) and 2900 settlement plots (SEMA, MT) with about 43.900 inhabitants in these areas (IBGE, 2011). Urban areas encompass 4000 ha and about 85.000 inhabitants (IBGE 2017b). The other 5 million ha have no registered tenure.

In terms of land use and tenure, indigenous lands play a key role in natural vegetation conservation - almost 37% of the forest and 22% of the cerrado are found in these areas. The Xingu National Park, encompassing 2,641,850 ha, or 15% of the basin area, is the largest patch of continuous forest in the landscape (Figure 2). Although large farms encompass the larger percentage of cerrado and the second largest percentage of forest, distribution throughout the landscape is patchy. Main activities are either pasture for cattle raising or single/double cropping (soybean and corn). Small and medium farmers have cleared almost only(?) natural vegetation and are producing either cattle or soybean. Most of the medium to large farmers use a conventional farming model, are highly skilled and in surveys(?) identified oil prices, logistics and manpower as the main constraints to increasing production.

Traditionally, soybean producers adopt Direct Seeding Mulch-Based Cropping System (soybean - corn) driven by market prices. Most of the crops are rain-fed, there are but a few irrigation systems due to high costs and the need for permits from the state. However, about 4000 small dams for cattle watering are found in the basin (Macedo et al. 2013).

About 50% of the soybean that is produced in the UXRB is exported to other countries. This trade results in a Gross Domestic Product (GDP) of €1,3 million for the region, with a trend of exponential growth. Another interesting pattern is the shifting amongst trading partners; while in the early 2000s Europe was the main buyer, Asia took this place over the course of the next decade, starting with China, Vietnam and Malaysia (Figure 3). High productive chain costs translate into pressure on farmers to negotiate the product at lower prices.

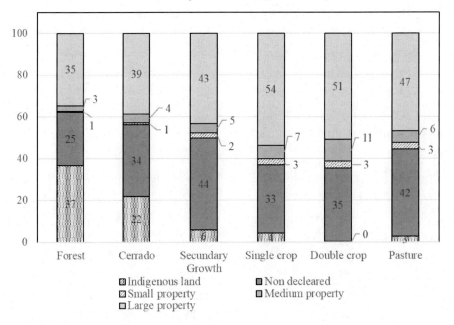

Figure 2. The Upper Xingu River Basin land use and land cover area (percentage) classified by type of stakeholder.

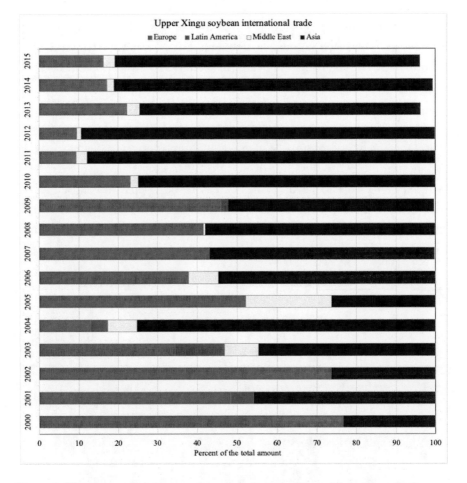

Figure 3. Time series of soybean main exports from the Upper Xingu River Basin (Mato Grosso, Brazil) to different continents.

Our modelled results (Figure 4) show that 2025 governmental productivity targets can be already achieved in the business as usual scenario (25% increase in productivity). This will require changes in land use and soil management, such as a restoration of degraded pastures, use of secondary growth areas for cattle ranging and an increase of 80% in double cropping area. While this scenario demands an increase of about 241300 Tons of fertilizer application and would use 39% more water, there is no need for new areas to be deforested, and potential carbon sequestration can increase by 25%. A similar pattern was found for conventional intensification, but with higher yields and ecosystem services use. In this scenario, productivity can increase ~39%, while nutrient demands are 28.8% higher than in the previous one. The main tradeoffs are an increase of almost 3% in carbon sequestration, while only 0.9% more water would be required, compared to the business as usual scenario. As expected, sustainable intensification was the better scenario, providing a potential increase of 49% in productivity and 43% in carbon sequestration. Water and nutrient demands decrease over time, requiring 23.5% less fertilizer and 32% less water. Since in all scenarios there was no need for new areas to be opened for agricultural expansion, biodiversity showed a small increase in the business as usual scenario, followed by larger increases in the conventional intensification and sustainable intensification scenarios.

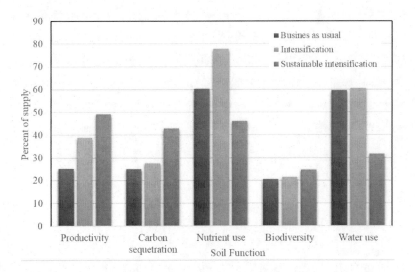

Figure 4. Soil function demand (as in Schulte and colleagues (2014)) for three scenarios (business as usual, conventional intensification and sustainable intensification) to meet Brazil's 2025 production targets in the Upper Xingu River Basin (Mato Grosso, Brazil).

Our results show that the functional land management approach is a powerful tool to understand and compare ecosystem services in multifunctional landscapes. This approach offers a simple methodology to compare different land management regarding demands and supplies of ES. Providing different scenarios to stakeholders to access tradeoffs between soil functions and ES degradation can facilitate the process of reconciling anthropogenic land use with the preservation and restoration of natural ecosystems and their services.

REFERENCES

AliceWeb2. 2016. *"System Of Analysis of Foreign Trade Information."* 2016. http://homologacao.investexportbrasil.gov.br/brazilian-statistics-and-economic-indicators.

Alves, Francisco Tarcísio, Carlos Frederico Lins E Silva Brandão, Kleybiana Dantas Da Rocha, Janaina Teixeira Da Silva, Luiz Carlos Marangon, and Rinaldo Luiz Caraciolo Ferreira. 2007. " Diametric and hypsometric structure of the arboreous component in fragments of atlantic rain forest in Recife-PE." *Cerne* 13 (1): 83–95.

Arvor, Damien, Margareth Meirelles, Vincent Dubreuil, Agnès Bégué, and Yosio E. Shimabukuro. 2012. "Analyzing the Agricultural Transition in Mato Grosso, Brazil, Using Satellite-Derived Indices." *Applied Geography* 32 (2): 702–13. https://doi.org/10.1016/j.apgeog.2011. 08.007.

Ballester, Maria Victoria Ramos, Reinaldo L Victoria, Alex V Krusche, M Bernardes, C Neill, and L A Deagan. 2013. "Physical and Human Controls on the Carbon Composition of Organic Matter in Tropical Rivers: An Integrated Analysis of Landscape Properties and River Isotopic Composition." In *Application of Isotope Techniques for Water Quality Assessment and Management, Focusing on Nutrient Management in Rivers*, edited by International Atomic Energy Agency, 173–86. Viena: IAEA.

Batistella, Mateus, and Emilio F Moran. 2007. "A Heterogeneidade Das Mudanças de Uso E Cobertura Das Terras Na Amazônia: Em Busca de Um Mapa Da Estrada." In *Dimensões Humanas Da Biosfera-Atmosfera Na Amazônia*, 65–80. São Paulo: EDUSP. [The Heterogeneity in Land Use and Land Cover Changes in the Amazon: Searching for a Road

Map. In: Human Dimensions of t0he Biosphere-Atmosphere in the Amazon, 65–80].

Brazil. 2012. *Brazilian Native Vegetation Protection Law.* http://www. planalto.gov.br/ccivil_03/_Ato2011-2014/2012/Lei/L12651.htm.

——. 2018. *Projeções Do Agronegócio*: Brasil 2017/18 a 2027/28 Projeções de Longo Prazo. Brasília: MAPA/ACE.

Castro Victoria, Daniel De, Alailson Venceslau Santiago, Maria Victoria Ramos Ballester, Antonio Roberto Pereira, Reynaldo Luiz Victoria, and Jeffrey E. Richey. 2007. "Water Balance for the Ji-Paraná River Basin, Western Amazon, Using a Simple Method through Geographical Information Systems and Remote Sensing." *Earth Interactions* 11 (5). https://doi.org/10.1175/EI198.1.

Coe, M. T., E. M. Latrubesse, M. E. Ferreira, and M. L. Amsler. 2011. "The Effects of Deforestation and Climate Variability on the Streamflow of the Araguaia River, Brazil." *Biogeochemistry* 105 (1): 119–31. https://doi.org/10.1007/s10533-011-9582-2.

Corbeels, Marc, Robelio Leandro Marchão, Marcos Siqueira Neto, Eliann Garcia Ferreira, Beata Emöke Madari, Eric Scopel, and Osmar Rodrigues Brito. 2016. "Evidence of Limited Carbon Sequestration in Soils under No-Tillage Systems in the Cerrado of Brazil." *Scientific Reports* 6. https://doi.org/10.1038/srep21450.

Coy, Martin, and Gerd Kohlhepp. 2005. *Amazônia Sustentável.* Rio de Janeiro: Garamond. [Sustainable Amazon].

Deegan, Linda A., Christopher Neill, Christie L. Haupert, M. Victoria R. Ballester, Alex V. Krusche, Reynaldo L. Victoria, Suzanne M. Thomas, and Emily de Moor. 2011. "Amazon Deforestation Alters Small Stream Structure, Nitrogen Biogeochemistry and Connectivity to Larger Rivers." *Biogeochemistry* 105 (1): 53–74. https://doi.org/10.1007/ s10533-010-9540-4.

EMBRAPA, Empresa Brasileira de Pesquisa Agropecuária. 2014. "*Banco de Dados de Solos Do Brasil.*" 2014. https://www.bdsolos.cnptia. embrapa.br/consulta_publica.html. [Brazil Soil Database].

ESRI. 2017. "*Environmental Systems Research Institute.*" 2017. https:// www.esri.com/.

FAO, Food and Agriculture Organization of the United Nations. 2019. *"FAOSTAT: Food and Agriculture Data."* 2019. www.fao.org/statistics/en/.

Foley, Jonathan A, Navin Ramankutty, Kate A. Brauman, Emily S. Cassidy, James S. Gerber, Matt Johnston, Nathaniel D. Mueller, et al. 2011. "Solutions for a Cultivated Planet." *Nature* 478 (7369): 337–42. https://doi.org/10.1038/nature10452.

FUNAI, Fundação Nacional do Índio. 2019. *"Coordenação Regional Do Xingu."* 2019. http://www.funai.gov.br/index.php/apresentacao-xingu. [Xingu Regional Coordination].

Garcia, Andrea S., Henrique O. Sawakuchi, and Maria Victoria R. Ferreira, Manuel Eduardo Ballester. 2017. "Landscape Changes in a Neotropical Forest-Savanna Ecotone Zone in Central Brazil: The Role of Protected Areas in the Maintenance of Native Vegetation." *Journal of Environmental Management* 187 (1): 16–23. https://doi.org/10.1016/j.jenvman.2016.11.010 0301-4797.

Garcia, Andrea S., Vívian M. de F. N. Vilela, Rodnei Rizzo, Paul West, James S. Gerber, Peder M. Engstrom, and Maria Victoria R. Ballester. 2019. "Assessing Land Use/cover Dynamics and Exploring Drivers in the Amazon's Arc of Deforestation through a Hierarchical, Multi-Scale and Multi-Temporal Classification Approach." *Remote Sensing Applications: Society and Environment* 15 (August): 100233. https://doi.org/10.1016/j.rsase.2019.05.002.

Heckenberger, Michael J., Afukaka Kuikuro, Urissapá Tabata Kuikuro, J. Christian Russell, Morgan Schmidt, Carlos Fausto, and Bruna Franchetto. 2003. "Amazonia 1492: Pristine Forest or Cultural Parkland?" *Science* 301 (5640): 1710–14. https://doi.org/10.1126/science.1086112.

Heckenberger, Michael J., J. Christian Russell, Carlos Fausto, Joshua R. Toney, Morgan J. Schmidt, Edithe Pereira, Bruna Franchetto, and Afukaka Kuikuro. 2008. "Pre-Columbian Urbanism, Anthropogenic Landscapes, and the Future of the Amazon." *Science* 321 (5893): 1214–17. https://doi.org/10.1126/science.1159769.

IBGE, Instituto Brasileiro de Geografia e Estatística. 1988. *"Mapa de Vegetação Do Brasil 1:5.000.000."* 1988. https://biblioteca.ibge. gov.br/biblioteca-catalogo.html?view=detalhes&id=66105.

———. 2017a. *"Censo Agropecuário."* 2017. https://censos.ibge.gov.br/ agro/2017/.

———. 2017b. *"Censo Demográfico."* 2017. https://www.ibge.gov.br/ estatisticas/multidominio/genero/9662-censo-demografico-2010.html? =&t=o-que-e. [Demographic Census].

INCRA, *Instituto Nacional de Reforma Agrária.* 2016. "Acervo Fundiário." 2016. http://acervofundiario.incra.gov.br/acervo/acv.php. [Land Collection].

Landmark. 2019. *"Landmark Glossary."* 2019. http://landmark2020.eu/ landmark-glossary/.

Leite, Christiane C., Marcos H. Costa, Cleverson A. de Lima, Carlos A.A.S. Ribeiro, and Gilberto C. Sediyama. 2011. "Historical Reconstruction of Land Use in the Brazilian Amazon (1940-1995)." *Journal of Land Use Science* 6 (1): 33–52. https://doi.org/10.1080/1747423X.2010.501157.

Macedo, Marcia N., Ruth S. DeFries, Douglas C. Morton, Claudia M. Stickler, Gillian L. Galford, and Yosio E. Shimabukuro. 2012. "Decoupling of Deforestation and Soy Production in the Southern Amazon during the Late 2000s." *Proceedings of the National Academy of Sciences of the United States of America* 109 (4): 1341–46. https://doi.org/10.1073/pnas.1111374109.

Macedo, Marcia N, Michael T Coe, R DeFries, M Uriarte, Paulo M Brando, C Neill, and W S Walker. 2013. "Land-Use-Driven Stream Warming in Southeastern Amazonia." *Philosophical Transactions of the Royal Society B: Biological Sciences* 368 (1619): 20120153–20120153. https://doi.org/10.1098/rstb.2012.0153.

Mello, Neli Aparecida de. 2006. *Políticas Territoriais Na Amazônia.* Annablume. [Territorial Policies In The Amazon].

Miranda, Eduardo, Janaina Carmo, Eduardo Couto, and Plínio Camargo. 2016. "Long-Term Changes in Soil Carbon Stocks in the Brazilian Cerrado Under Commercial Soybean." *Land Degradation and Development* 27 (6): 1586–94. https://doi.org/10.1002/ldr.2473.

MMA, Ministério do Meio Ambiente. 2015. "*Geo Catálogo.*" 2015. http://geocatalogo.mma.gov.br/. [Geo Catalog].

Morton, Douglas C, Ruth S DeFries, Yosio E Shimabukuro, Liana O Anderson, Egidio Arai, Fernando del Bon Espirito-Santo, Ramon Freitas, and Jeff Morisette. 2006. "Cropland Expansion Changes Deforestation Dynamics in the Southern Brazilian Amazon." *Proceedings of the National Academy of Sciences of the United States of America* 103 (39): 14637–41. https://doi.org/10.1073/pnas.060 6377103.

O'Sullivan, Lilian, David Wall, Rachel Creamer, Francesca Bampa, and Rogier P.O. Schulte. 2018. "Functional Land Management: Bridging the Think-Do-Gap Using a Multi-Stakeholder Science Policy Interface." *Ambio* 47 (2): 216–30. https://doi.org/10.1007/s13280-017-0983-x.

Pielke, Roger A. 2005. "Misdefining 'climate Change': Consequences for Science and Action." *Environmental Science and Policy* 8 (6): 548–61. https://doi.org/10.1016/j.envsci.2005.06.013.

Raucci, Guilherme Silva, Cindy Silva Moreira, Priscila Aparecida Alves, Francisco F.C. Mello, Leidivan De Almeida Frazão, Carlos Eduardo P. Cerri, and Carlos Clemente Cerri. 2015. "Greenhouse Gas Assessment of Brazilian Soybean Production: A Case Study of Mato Grosso State." *Journal of Cleaner Production* 96: 418–25. https://doi.org/10.1016/j. jclepro.2014.02.064.

Rausch, Lisa L., and Holly K. Gibbs. 2016. "Property Arrangements and Soy Governance in the Brazilian State of Mato Grosso: Implications for Deforestation-Free Production." *Land* 5 (2). https://doi.org/10. 3390/land5020007.

Riskin, Shelby H., Stephen Porder, Christopher Neill, Adelaine Michela e Silva Figueira, Carmen Tubbesing, and Natalie Mahowald. 2013. "The Fate of Phosphorus Fertilizer in Amazon Soya Bean Fields." *Philosophical Transactions of the Royal Society B: Biological Sciences* 368 (1619). https://doi.org/10.1098/rstb.2012.0154.

Rizzo, Rodnei, Andrea S. Garcia, Maria Victoria R. Ballester, Daniel C. Victoria, Christopher Neill, and Humberto R. da Rocha. 2019. "Land Use Changes and Alteration in Water Resources of Southeastern

Amazon: Trends in Rainfall, Temperature and River Discharge during 1976-2015." *Climate Change* (in Review).

Santos, H. G., W. Carvalho Júnior, R. O. Dart, M. L. D. Áglio, J. S. Sousa, J. G. Pares, A. Fontana, A. L. S. Martins, and A. P. O. Oliveira. 2011. "O Novo Mapa de Solos Do Brasil: Legenda Atualizada." *Embrapa Solos*, 67. https://www.embrapa.br/busca-de-publicacoes/-/publicacao/920267/o-novo-mapa-de-solos-do-brasil-legenda-atualizada. [The New Soil Map of Brazil: Updated Classification].

Sawakuchi, Henrique O., Maria Victoria, Ballester, and Manuel Eduardo Ferreira. 2013. "The Role of Physical and Political Factors on the Conservation of Native Vegetation in the Brazilian Forest-Savanna Ecotone." *Open Journal of Forestry* 3 (1): 49–56. https://doi.org/10.4236/ojf.2013.31008.

Schaden, Egon. 1990. "Pioneiros Alemães Da Exploração Etnológica Do Alto Xingu." *Revista de Antropologia* (São Paulo) 33: 1–18. https://doi.org/10.11606/2179-0892.ra.1990.111211. [German Pioneers of Upper Xingu Ethnological Exploration].

Schulte, Rogier P. O., Rachel E. Creamer, Trevor Donnellan, Niall Farrelly, Reamonn Fealy, Cathal O'Donoghue, and Daire O'hUallachain. 2014. "Functional Land Management: A Framework for Managing Soil-Based Ecosystem Services for the Sustainable Intensification of Agriculture." *Environmental Science and Policy* 38: 45–58. https://doi.org/10.1016/j.envsci.2013.10.002.

SEMA-MT, Secretaria do Meio Ambiente do Estado do Mato Grosso. 2016. "*Sislam - Sistema Integrado de Monitoramento E Licenciamento Ambiental.*" 2016. https://monitoramento.sema.mt.gov.br/simlam/. [Integrated Environmental Monitoring And Licensing System].

USGS. 2019. "Climate Hazards Group InfraRed Precipitation with Station Data - Annual Precipitation." *CHIRPS*. 2019. https://earlywarning.usgs.gov/fews/datadownloads/Global/CHIRPS 2.0.

Velasquez, C., Alves, H. Q., Bernasconi, P. 2010. "*Fique Por Dentro*: *A Bacia Do Rio Xingu Em Mato Grosso*". São Paulo: Instituto Socioambiental, Instituto Centro de Vida. [*Be informed: The Xingu River Basin in Mato Grosso*].

In: Land Use Changes
Editor: Vinícius Santos Alves

ISBN: 978-1-53617-032-0
© 2020 Nova Science Publishers, Inc.

Chapter 4

BEGINNING OF DESERTIFICATION IN THE SOUTHERN BUENOS AIRES AND THE PREDICTABILITY OF SOIL WATER CONTENT

Luciana Stoll Villarreal[1], Marcela Hebe González[1,2], Alfredo Luis Rolla[2] and María Elizabeth Castañeda[1,2,]*

[1]Departamento de Ciencias de la Atmósfera y los Océanos, Facultad de Ciencias Exactas y Naturales, Universidad de Buenos Aires, Buenos Aires, Argentina

[2]Centro de Investigaciones del Mar y la Atmósfera, CONICET-UBA, Buenos Aires, Argentina

ABSTRACT

The implementation of seasonal forecasts of soil water in relative small spatial scales is of great interest, especially in the agricultural sector as they facilitate decision-making what allows a better management of

---

* Corresponding Author's Email: lucianastoll92@gmail.com.

water resources and maximize efficiency in productivity. In this work, Tres Arroyos meteorological station was chosen for the analysis. Located in the south of Buenos Aires, Tres Arroyos is one of the most important regions for corn production. The aim of this study is to design a statistical model using atmospheric forcing to predict soil water storage (WS) for spring. The analysis of the efficiency of different models takes into account the adjusted squared correlation coefficient ($\overline{R^2}$) and cross-validation coefficient (CV) values. The preliminary results show that the best designed model has an efficiency of around 66%.

Keywords: seasonal forecast, soil water, climate forcing

1. INTRODUCTION

1.1. Climate and Soil Situation in the Region

The water content in the soil depends mainly on temperature and precipitation. The global precipitation in the middle latitudes has decreased between 3 and 8 mm per year since 1951 and the temperature showed an increase of about 0.5°C (IPCC, 2013). In Argentina, an increase in average temperature and precipitation has been observed in most of its territory (Berbery et al., 2006, Barros et al., 2008, Barros et al., 2014, Saurral et al., 2016). However, there are regions where there are negative trends in precipitation. One of them is the northern Patagonian region (between 36°S and 43°S), specifically in the high mountain area in the west (Castañeda and González 2008, González and Vera 2010). The Third Communication of the Argentine Republic to the United Nations Framework Convention on Climate Change (UNCCF, 3CN, 2015) showed an increase in the average annual temperature of more than 0.5°C in the transition zone between the humid Pampas and Patagonia, while there was no significant change in annual precipitation. However, some localized regions registered a decrease in precipitation as the area of the mid-latitude Andes between 1960 and 2010 with serious consequences in an area where water availability is a critical factor for rainfed agriculture.

Future scenarios (RCP4.5 and RCP8.5) show an increase in the average annual temperature of between 0.5°C-1°C in the near future (2040) while the changes in precipitation are less significant, except for the Patagonian northwest region. Therefore, the transition zone between the Pampa plain and Patagonia will probably be affected by an increase in temperature, a slight decrease in precipitation, which will generate a lower contribution of water to the soil, which will seriously affect agricultural regional economies.

The soil water availability constitutes a primary factor in the different stages of crops growth, particularly, the soil water deficit during the flowering stage could considerably affect the yield (Denmead and Shaw, 1960; Sudar et al., 1981). In the southern Buenos Aires province, the department of Tres Arroyos, lat: 38.33°S; lon: 60.25°W (Figure 1) constitutes one of the main crop production regions. The climate of this region is temperate and sub-wet and is highly influenced by the Atlantic Ocean (Bohn et al., 2011, Campo et al., 2004) with high wind intensity (mean annual speed of 13.4 km/h with greater values in spring and summer; maximum value around 23 km/h in the period 1970-2007). Maximum values of precipitation are observed in autumn and spring with dry season in winter. Mean annual precipitation is 846.5mm considering the period 1979-2015 (Figure 2).

Observed mean annual temperature in the period 1970-2016 shows a significant positive trend (r = 0.4) (Figure 3a.) and this is even greater if the period 2001-2016 is considered (Figure 3b). Observed annual precipitation in the period 1979-2016 (Figure 3c.) and 2001-2016 (Figure 3d.) present a no significant negative trend.

This region presents sandy and dunes, very susceptible to wind and water erosion. On the other hand, the intervention of man has generated a decrease in Normalized Difference Vegetation Index (NVDI) (Gaitán et al., 2015) which indicates a lower amount of vegetation, soil compaction and loss of organic matter. This generates soil degradation and less productivity, which is why the region presents a clear process of beginning of desertification (Pérez Pardo, 2002).

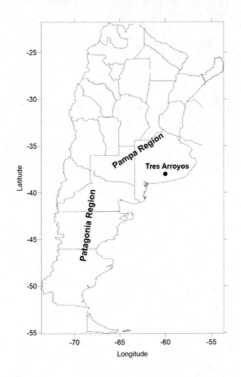

Figure 1. Location of Tres Arroyos, the meteorological station in study. Pampa region and Patagonia Region also are shown.

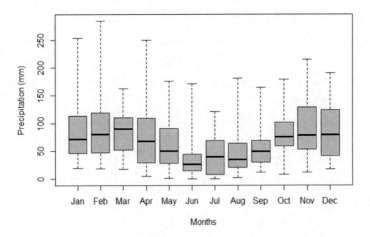

Figure 2. Monthly precipitation in Tres Arroyos for the period 1979-2015. Whiskers show the minimum and maximum values; the box shows the first and third quartiles; the line shows the median.

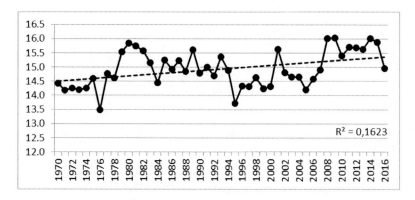

Figure 3a. Mean annual temperature (°C) in the period 1970-2016 (solid line) and linear trend (dashed line).

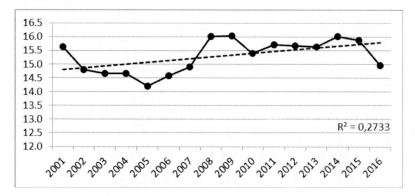

Figure 3b. As in Figure 3a, but in the period 2001-2016.

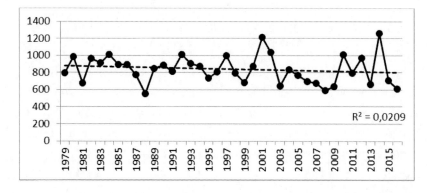

Figure 3c. Annual precipitation (mm) in the period 1979-2016 (solid line) and linear trend (dashed line).

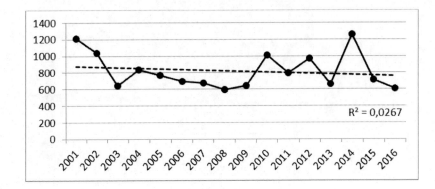

Figure 3d. As in Figure 3c, but in the period 2001-2016.

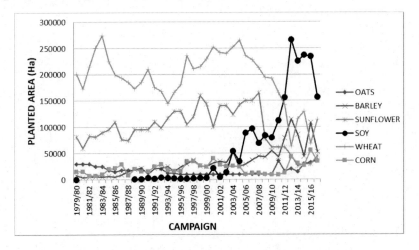

Figure 4. Main crops in Tres Arroyos, according to the planted area (in hectares) in the period 1979-2016. Source: Ministry of Agriculture, Livestock and Fisheries of the Nation (https://datos.agroindustria.gob.ar/dataset/estimaciones-agricolas).

This area presents a field capacity (FC) of 190-220mm and a wilting point (WP) between 95 and 100mm (Fernandez Long et al., 2012). Like many other crops in Argentinian Plain region, the wheat and corn develop mostly in rainfed conditions, and therefore, soil water plays a fundamental role and is strongly associated with crop production. Among the main crop species that are planted in the region are: wheat, corn, soy, sunflower, barley and oats according to the planted area (Figure 4). The usual planting date is May for wheat and October for corn in Tres Arroyos (Rolla et al., 2017). Fernandez Long et al., (2012) defined a Water Satisfaction Index (WSI) which depended on the soil water and showed that crop yields are highly related to WSI in some seasons of the year. The correlation between wheat yield and WSI was highly significant (almost 0.4) in October in Tres Arroyos. Therefore, it is important to provide soil water forecasts in October to improve cereal production.

Some authors studied that several climatic forcings affect seasonal precipitation and temperature (some of the parameters of water balance), particularly El Niño-Southern Oscillation (Compagnucci and Vargas, 1998, Grimm et al., 2002, Vera et al., 2004, Barreiro, 2010, Garbarini et al., 2016, among others). However, other climate patterns as for example The Antarctic Oscillation (AAO) (Thompson and Wallace, 2000) or the sea surface temperature in the Atlantic ocean near Argentina (Oliveri et al., 2018), among many others, influence the interannual variability of temperature and rainfall. Therefore, the circulation and sea surface temperature anomalies can affect the soil water content.

The aim of this work is to design a statistical model using previous meteorological variables to predict the WS in October (Stoll Villarreal et al., 2018). These predictions could help decision-makers to take accurate measures with some time in advance to mitigate the negative effect in periods with low water availability.

2. Methodology and Data

2.1. Description of Data

Meteorological data from Tres Arroyos were provided by the National Weather Service of Argentina. The record used in this study is 1979-2016. WS data (used to design the statistical model) were obtained from a model of water balance (Fernandez Long et al., 2012). There are not WS measurements and so, the WS derived from this model was used as the estimation closest to the true data. They will be called "MWS" to distinguish from the predictions derived from the statistical designed model ("PWS").

The water balance (MWS) proposed presents as main variables the potential evapotranspiration (ETP) (amount of water that the atmosphere could demand), real evapotranspiration (ETR) (actual loss of water by evapotranspiration depending on soil coverage), precipitation (PP) and change in water storage. ETP was calculated using Penman-Monteith equation following FAO recommendations (Allen et al., 1998). This model incorporates some modifications to the one proposed by Penman-Monteith, considering the soil characteristics (Forte Lay et al., 1995). These modifications tend to better represent water content in heavy soils (where WP is more than 40% of FC) since they tend to show drying in excess.

Water balance equation to determine WS is represented as follows:

WS = PP-ETR-EXC   (1)

where EXC is the excess of water, produced when WS exceeds FC. EXC gets lost by surface runoff and percolation.

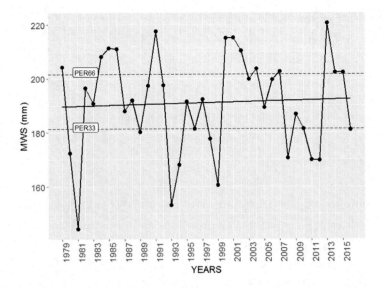

Figure 5. MWS evolution (1979-2016). Dashed horizontal lines show first and second terciles values of the distribution. Solid black line represents the fitted regression over data.

Table 1. Some parameters of MWS distribution (in mm)

Period	Mean	Standard Deviation	First Tercile	Second Tercile
1979-2016	190.96	18.43	181.45	201.63

MWS in October data are detailed in Figure 5. The trend of this series was calculated using a linear approximation (ordinary least squares) and the significance (95% of confidence) was tested using a Student's t-test. No significant linear trend was detected over the studied period. Some parameters (in mm) that characterize distribution of MWS in October are shown in Table 1. The first and second terciles were calculated because they represent useful thresholds to define extreme values.

Monthly reanalysis data from the European Centre (Era-Interim) of the ECMWF were used, with spatial resolution of 0.5° latitude by 0.5° longitude and global coverage (Dee et al., 2011) for 1979-2016 period. The meteorological variables considered to define possible predictors of WS were: geopotential height in 200, 500 and 1000 hPa (hgt200, hgt500, hgt1000), zonal and meridional wind at low levels (850 hPa) (u850, v850), temperature in 1000 hPa (t1000), precipitable water all along the atmospheric column (tcw), specific humidity in 850hPa and 1000hPa (q850, q1000), sea surface temperature (sst) and volume of soil water in two different levels, between 0 and 7cm deep (vls1) and between 7 and 28 cm deep (vsl2).

2.2. Methodology

Meteorological variables detailed above, averaged in September (S), in August-September bimester (AS) and in July-August-September quarter (JAS) were correlated with MWS in the following October. Some predictors could be defined as the mean value of these variables in the area where the correlation coefficients R were higher than 0.33 (significant with a confidence level of 95% using a normal test).

Then, it was necessary to establish a criterion of selection for the best predictors. The set of predictors that enter the model must be independent each other to avoid multicollinearity and only predictors with physical meaning were considered. Multiple linear regression with backward selection method was applied over different sets of predictors to elaborate the forecast models.

The best models were selected using two criterions: they would explain the greatest possible MWS variance (the greatest adjusted $\overline{R}^2$) and they would be stable (the lowest cross-validation coefficient) (Hyndman and Athanasopoulos, 2013).

$\overline{R}^2$ is calculated as:

$$\overline{R}^2 = 1 - (1 - R^2)\frac{T-1}{T-K-1} \tag{2}$$

where T is the number of years, K is the number of predictors and R the linear correlation coefficient. This is an improvement of R^2, as it will not increase with each added predictor.

Cross-validation coefficient (CV) is calculated with a method in which available data are repeatedly divided into developmental and verification data subsets (Wilks, 1995). This procedure consists in evaluating the performance of a forecast equation constructed with part of the data, and using the rest of the data as the verification set. Although exist several ways to calculate CV, in this work the leave-one-out method was utilized, so different forecasts equations were calculated using all possible combinations of partitions. Then, the difference between MWS (y_t) and WS forecast ($\hat{y}_t$) for each year (t) is found. CV is the average of the mean square error, it is computed as:

$$CV = \frac{1}{N}\Sigma(y_t - \hat{y}_t)^2 \tag{3}$$

where N is the number of years analyzed. This method is used because the record is too short to divide the total period in one for training and other for verification and so it is a good way to prove not only the model efficiency but the stability of the model. Therefore, models with minimum CV and $\overline{R}^2$ higher than 0.44 (explaining more than 44% of the MWS variance) were chosen.

Both PWS (resulting from cross-validation method) and MWS values were classified in three categories according to the terciles of the MWS distribution: below normal (BN) values less than first tercile, normal (N) values between first and second terciles and above normal (AN) values greater than second tercile. First MWS tercile is 181.45 mm and second tercile is 201.63 mm. Therefore, it was possible to know the efficiency of each obtained model comparing the categories of MWS and PWS series.

Other measures of accuracy were calculated based on contingency tables for each category (BN, N and AN) separately. The indices considered were: Hit Rate (HR), Probability of Detection (POD) and False-Alarm Rate (FAR) (Wilks, 1995). HR index gives the proportion of events that were correctly forecasted, POD index indicates the probability of forecast an event when it is observed and finally FAR is the proportion of forecasted events that fail to occur. The best HR and POD values are 1 meanwhile the best FAR value is 0.

Receiver-Operating-Characteristic (ROC) curves were also incorporated to the analysis of efficiency. This method consists in represent the relationship between sensitivity and specificity of a model. Sensitivity is defined as the probability of correctly classify an event that occurs and takes the same value of POD, defined above. Specificity, meanwhile, is the probability of forecasting that an event will not occur when it does not really occur. Generally, sensitivity and specificity (which take values between 0 and 1), are plotted in a same space called ROC space. The x-axis represents 1-specificity and y-axis, sensitivity. Therefore, models with highest values of sensitivity and specificity will be considered the best. Frequently, a measure used to quantify the accuracy of a model is the area under the ROC curve (AUC), calculated as:

$$AUC = 0.5 + \frac{|(1-\text{specificity})-\text{sensitivity}|}{2} \qquad (4)$$

Values of AUC near to 1 indicate that the model has high specificity and sensitivity, so it well discriminate between different groups or categories.

3. RESULTS

3.1. The Statistical Models

The correlation fields between MWS in October and previous meteorological variables were used to define some predictors independent each other and with physical meaning. Predictors that better described MWS are summarized in Figure 6.

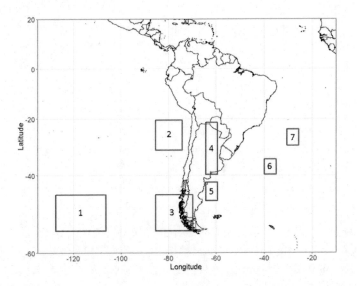

Figure 6. Regions 1 to 7 represent the location of predictors. AShgt200$_1$ (1), AShgt200$_2$ (2), Su850 (3), Sq850 (4), JASsst (5), Shgt500 (6), ASsst (7).

AShgt200$_1$ is the mean hgt200 in August-September in South Pacific and it represents the position of polar jet. It is the result of the energy transport derived from transient systems. Convergence of the equatorial side at the exit of the jet and the polar side at the entrance, and divergence of the polar side at the exit and from the equatorial side at the entrance are associated with this jet. Areas with divergence at high levels are related to convergence at lower levels and the associated convection and storms (Carlson, 1998). The predictor AShgt200$_2$ is the mean hgt200 in August-September on the Pacific coast near the north of Argentina and represents the possible action of the subtropical jet. At low levels, the mean u850 in September in South Pacific and Patagonia (Su850) is located in westerlies area in southern Argentina. This is the area where the Antarctic Oscillation (AAO) is defined (Thompson and Wallace, 2000). The positive (negative) phase of AAO is associated with lower (higher) energetic exchange between latitudes and lower (greater) possibility of fronts displacement associated with storm systems that move towards the northeast (Reboita et al., 2009; Silvestri and Vera, 2003).

The jet stream and the Rossby waves displacing over the Pacific Ocean constitute a typical pattern of Southern Hemisphere related to precipitation systems entering Argentina across southern Andes and then displacing towards the northeast over the continent (Kidson, 1999; Mo, 2000; Paegle and Mo, 2002).

Sq850 is the mean q850 in September averaged in central and northern Argentina, west Paraguay and South-East Bolivia. It is associated with the humidity transport from the Brazilian forest through the low level jet, a typical spring and summer pattern in South America (Satyamurty et al., 1998; Salio et al., 2002; Vera et al., 2006). Mean SST in July-August-September (JASsst) in South Atlantic Ocean and Patagonian coast is related to humidity and heat advection, both with especial influence over temperature (Oliveri, 2018).

Shgt500 is the mean hgt500 in September and ASsst is the mean SST in August-September over the South Atlantic Ocean. Both predictors are associated with the entrance of air through the semi-permanent Atlantic High. Particularly, ASsst can be related to the phase of South Atlantic Dipole (SAODI) defined by Nnamchi et al. (2011) (González et al., 2015; González et al., 2017; Sun et al., 2017) and to the presence of the South Atlantic Convergence Zone (Satyamurty et al., 1998; Barros et al., 2000). Although this is a typical summer feature, sometimes it is present in spring. All these predictors are highly correlated with MWS (significant at 95% confidence level).

Four sets of independent predictors could be defined (Table 2). Each one was entered the multiple linear regression using backward methodology and the best derived model from each set is detailed in the first column where the subscripts indicate the regions where predictors were defined. The $\overline{R}^2$ and CV values of the models are also detailed in Table 2. The MWS and PWS with different models are detailed in Figures 7a. (model 1), 7b. (model 2), 7c. (model 3), 7d. (model 4) and 7e. (ensemble). It is important to note that in all cases the series derived from the models are very similar to that resulting from the cross-validation method, indicating the stability of those models.

Table 2. PWS equations derived from statistical models (1 to 4). Sets of predictors incorporated to the models are detailed. Also CV and adjusted R^2 values for each model are shown

	Statistical model	Set of predictors	CV	Adjusted R2
1	PWS=6559.6(Sq850)-0.01(AShgt2001)+0.008(AShgt2002)+23.88(ASsst)-12.79(JASsst)-2588.2	Sq850 AShgt2001 AShgt2002 ASsst JASsst	212.49	0.474
2	PWS=8253.52(Sq850)-0.02(AShgt2002)+23.47(ASsst)-12.9(JASsst)-1303.14	Sq850 AShgt2002 ASsst JASsst	214.13	0.455
3	PWS=7155.71(Sq850)+0.004(Shgt500)+0.009(AShgt2002)+21.46(ASsst)-14.21(JASsst)-3372.6	Sq850 Shgt500 AShgt2002 ASsst JASsst	215.67	0.447
4	PWS=-1.935(Su850)+0.008(Shgt500)+0.012(AShgt2002)+18.41(ASsst)-13.36(JASsst)-3220.1	Su850 Shgt500 AShgt2002 ASsst JASsst	215.99	0.438

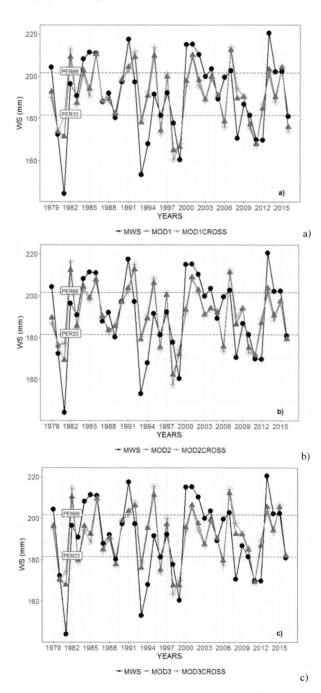

Figure 7. (Continued).

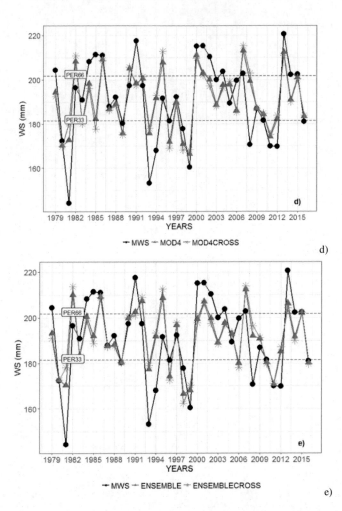

Figure 7. Time series of WS in the period 1979-2016 according to MWS and PWS using: a) model 1 and model 1 after cross-validation methodology, b) model 2 and model 2 after cross-validation methodology, c) model 3 and model 3 after cross-validation methodology, d) model 4 and model 4 after cross-validation methodology and e) ensemble of the models 1 to 4 and ensemble after cross-validation metodology. First and second terciles in dashed horizontal lines.

All the models explain more than 43% of the MWS variance. Model 1 explains 47.4% of MWS variance and has the smallest CV (212.49) showing the best performance.

3.2. Model Accuracy

Efficiency coefficients were calculated using categories (BN, N and AN) for MWS and PWS derived from each one of the models and the ensemble using cross-validation series.

The measures of accuracy detailed in section 2.2, are shown in Table 3. Model 1 has the highest efficiency in all categories, this is because HR and POD indices are the nearest to 100%, whereas FAR index takes the lowest value (0%). Even if we consider the possibility of detecting the deficit of soil water as the main objective, probably POD is the most relevant parameter and therefore model 1 achieve the best performance with 77% of BN events detected and only 10% of FAR.

Table 3. Accuracy measures for each model and for the ensemble.
Values are presented for Below Normal, Normal and Above
Normal events

Category	Model	HR	POD	FAR	Specificity	AUC
BN	1	89	77	10	96	0.87
	2	87	69	10	96	0.83
	3	84	69	18	92	0.81
	4	82	62	20	92	0.77
	ENS	89	77	10	96	0.87
	1	66	67	53	65	0.66
	2	61	58	59	62	0.60
N	3	58	58	61	58	0.58
	4	58	58	61	58	0.50
	ENS	61	67	58	54	0.61
	1	76	54	30	88	0.71
	2	74	54	36	84	0.69
AN	3	74	46	33	88	0.67
	4	66	38	50	80	0.59
	ENS	71	31	38	88	0.60

The ROC space and pairs (1-E, S) for each model is presented in figure 8. BN events were the best forecasted category, followed by AN category.

However, normal events are not detected properly by any model. The area below curve (AUC) detailed in Table 3 shows that model 1 seems to be the best of all as it has the highest AUC in all the categories.

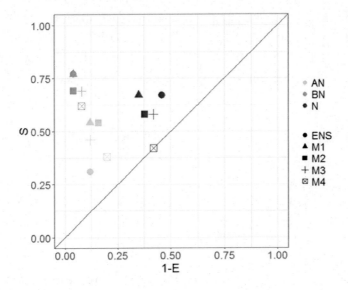

Figure 8. Sensitivity vs (1-specificity) for all the models (M1 to M4) and the ensemble (ENS) using categorized forecast (BN, N, AN).

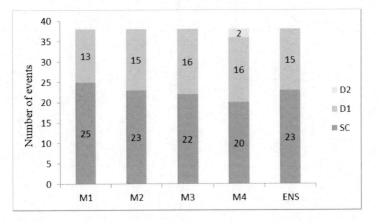

Figure 9. Number of events of PWS that were classified in the same category (SC) that MWS, that differed in one category (D1) and that differed in 2 categories (D2) for each model (M1 to M4) and the ensemble (ENS).

Figure 9 shows the comparison between PWS and MWS classification. The results show that the ensemble has an effectivity of around 61% (out of 38 events, 23 were classified in the right category whereas 15 events differ only in one category). The model 1 achieved the highest efficiency, around 66% of events were properly classified.

CONCLUSION

Multiple linear regression technique was used to estimate WS in Tres Arroyos in spring. The results indicate that the best predictors are specific humidity over central and northern Argentina, zonal wind in 850 hPa in Patagonia and South Pacific Ocean, 200 hPa geopotential height over South Pacific Ocean, 500 hPa geopotential height and sea surface temperature over South Atlantic Ocean.

The ensemble of the best four selected models produced good results, it showed an efficiency of about 61%, and so 6 out of 10 events were classified properly. Particularly BN category showed 77% of probability of detection.

These results are encouraging taking into account that the focus of this work is to anticipate events with soil water deficit. However, these forecast models can be improved to detect extreme values. The methodology used to build the models could influence the accuracy of the models. So as future work, other methodologies such as non-linear models, autoregressive and neural networks will be tested.

ACKNOWLEDGMENTS

The authors thank Mg. Maria Elena Fernandez Long and Ing. Liliana Spescha for providing soil water data. To the National Weather Service in Argentina for providing meteorological data.

European Centre for Medium-range Weather Forecast (ECMWF) (2011): The ERA-Interim reanalysis dataset, Copernicus Climate Change

Service (C3S), available from (https://www.ecmwf.int/en/forecasts/datasets/archive-datasets/reanalysis-datasets/era-interim).

Ministry of Agriculture, Livestock and Fisheries of the Nation for the productivity data obtained from its webpage (https://datos.agroindustria.gob.ar/dataset/estimaciones-agricolas).

This research was supported by UBACYT 2017-2019200201601000009BA, UBACYT Interdisciplinary 2018-2020 20620170100012BA and CONICET PIP 2015-2017 projects.

REFERENCES

Allen, R. G., Pereira, L. S., Raes, D. and Smith, M. (1998). Crop evapotranspiration – Guidelines for computing crop water requirements– FAO irrigation and drainage paper 56, Food and Agriculture Organization of the United Nations. Rome, 1998.

Barreiro, M. (2010). Influence of ENSO and The South Atlantic Ocean on climate predictability over Southeastern South America. *ClimDyn*. 35, 1493-1508.

Barros, V., González, M. H., Liebmann, B. and Camilloni, I. (2000). Influence of the South Atlantic Convergence Zone and South Atlantic sea surface temperature on interannual summer rainfall variability in southeastern South America. *Theoretical and Applied Climatology*, ed. Springer, vol 67, n° 3 y 4, 123-133 pp. ISSN 0177 798X.

Barros, V., Doyle, M. and Camilloni, I. 2008. Precipitation trends in southeastern South America: relationship with ENSO phases and the low-level circulation. *Theoretical and Appl. Climatology*. 93, 1-2: 19-33.

Barros, V., Boninsegna, J. A., Camilloni, I., Chidiak, M., Magrín, G. O. and Matilde Rusticucci. (2014). Climate change in Argentina: trends, projections, impacts and adaptation. *WIREs Clim Change* 2014. doi: 10.1002/wcc.316.

Berbery, E., Doyle, M y Barros, V, 2006. Las tendencias regionales de precipitación. Capítulo del libro "El cambio climático en la Cuenca del

Plata," Barros, Clarke y Silva Dias (editores). [Regional precipitation trends. Chapter in *Climate change in the Plata Basin*].

Bohn, V. Y., Piccolo, M. C., Perillo, G. M. E. (2011). Análisis de los periodos secos y húmedos en el sudoeste de la provincia de Buenos Aires (Argentina). Agencia Estatal de España. Revista de Climatología; 11; 3-2011; 31-43. [Analysis of dry and wet periods in the southwest of Buenos Aires province (Argentina). State Agency of Spain. *Climatology journal*; 11; 3-2011; 31-43.

Campo, A., Capelli, A. and Diez, P. (2004). El clima del Sudoeste Bonaerense. EdiUns, Bahía Blanca. [Southwest of Buenos Aires province *Climate. EdiUns, Bahía Blanca*].

Carlson, T. N. (1998). *Mid-latitude weather systems*. American Meteorological Society.

Castañeda, E. y González, M. 2008. Some aspects related to precipitation variability in the Patagonia region in Southern South America. *Atmósfera*. 21(3), 303-317.

Compagnucci, R. H. and Vargas, W. (1998). Interannual variability of Cuyo Rivers streamflow in Argentinean Andean Mountains and ENSO events. *Int. J. Climatol*. 18, 1593–1609.

Dee, D. P., Uppala, S. M., Simmons, A. J., Berrisford, P., Poli, P., Kobayashi, S., Andrae, U., Balmaseda, M. A., Balsamo, G., Bauer, P., Bechtold, P., Beljaars, A. C. M., van de Berg, L., Bidlot, J., Bormann, N., Delsol, C., Dragani, R., Fuentes, M., Geer, A. J., Haimberger, L., Healy, S. B., Hersbach, H., Holm, E. V., Isaksen, L., Kållberg, P., Köhler, M., Matricardi, M., McNally, A. P., Monge-Sanz, B. M., Morcrette, J-J, Park, B-K, Peubey, C., de Rosnay, P., Tavolato, C., Thépaut, J-N and Vitart, F. (2011). The ERA-Interim reanalysis: configuration and performance of the data assimilation system. *Q. J. R. Meteorol. Soc*. 137: 553–597.

Denmead, O. T. and Shaw, R. H. (1960). The effects of soil moisture stress at different stages of growth on the development and yield of corn. *Agron. J*. 52: 272-274.

Fernández Long, M. E., Spescha, L., Barnatán, I. and Murphy, G. (2012). Modelo de Balance Hidrológico Operativo para el Agro (BHOA). Rev.

Agronomía & Ambiente 32(1-2): 31-47. FA-UBA, Buenos Aires, Argentina. [Operational Hydrological Balance Model for Agro (BHOA). *Rev. Agronomy & Environment* 32 (1-2): 31-47. Faculty of Agronomy-University of Buenos Aires].

Forte Lay, J. A., Aiello J. L. and Kuba, J. (1995). *Agroagua Software,* version 4.0, preprint Agrosoft 95, Juiz de Fora, Brasil.

Gaitán, J. J., Donaldo Bran D. E and Azcona, C. (2015). Trend of NDVI in the period 2000-2014 as indicator of land degradation in Argentina: advantages and limitations. *Agriscientia* 32 (2): 83-93.

Garbarini, E. M., Skansi, M., González, M. H. and Rolla A. (2016). ENSO Influence over Precipitation in Argentina, *Advances in Environmental Research*, Chapter 7, Volume 52, NOVA Publisher, NY, USA.

González, M. H., Garbarini, E. M., Rolla, A. L. and Eslamian, S. (2017). Meteorological Drought Indices: Rainfall Prediction in Argentina en Handbook of Drought and Water Scarcity: Vol. 1, *Principle of Drought and Water Scarcity,* Chapter 29, 540-567. Taylor & Francis Publishing (CRC Group) Editor: Saeid Eslamian. ISBN: 9781498731089 1498731082. Reino Unido, Abingdon.

González, M. H., Garbarini, E. M. and Romero P. E. (2015). Rainfall patterns and the relation to atmospheric circulation in northern Patagonia (Argentina), *Advances in Environmental Research*, Chapter 6, 41, 85-100, Editors: Justin A. Daniels, NOVA Publisher, NY, USA ISBN 978-1-63482-885-7

González, M. H. and Vera, C. S. (2010). Interannual winter rainfall variability in Southern Andes. *International Journal of Climatology.* ISSN 0899 8418. RMS, Reading, Reino Unido. DOI: 10.1002/joc.1910.

Grimm, A., Barros, V. and Doyle, M. (2002). Climate variability in Southern South America associated with El Niño and La Niña events. *J. Climate*, 13, 35-58.

Hyndman, R. J. and Athanasopoulos, G. (2013). *Forecasting: principles and practice.* OTexts. 292 pages (https://otexts.org/fpp2/selecting-predictors.html).

IPCC 2013. Cambio Climático 2013, Bases físicas, Contribución del Grupo de trabajo I al Quinto, Informe de Evaluación del Grupo

Intergubernamental de Expertos sobre el Cambio Climático. [*Climate Change 2013*, Physical basis, Contribution of Working Group I to the Fifth, Evaluation Report of the Intergovernmental Panel on Climate Change.]

Kidson, J. (1999). Principal modes of southern hemisphere low frequency variability obtained from NCEP-NCAR reanalyses. *J. Climate* 1:1177-1198.

Mo, K. C. (2000). Relationships between low frequency variability in the Southern Hemisphere and sea surface temperature anomalies. *J. Climate* 13: 3599-3610.

Nnamchi, H. C., Li, J. and Anyadike, R. (2011). Does a dipole mode really exist in the South Atlantic Ocean? *J. Geophys. Res.* 116, doi: 10.1029/2010JD015579.

Nnamchi, H. C. and Li, J. (2011). Influence of the South Atlantic Ocean Dipole on West African summer precipitation. *J. Climate* 24: 1184-1197.

Oliveri, P. (2018). La influencia de los océanos cercanos sobre la precipitación y temperatura media estacionales en Argentina. Tesis de Licenciatura. Departamento de Ciencias de la Atmósfera y los Océanos, Facultad de Ciencias Exactas y Naturales, Universidad de Buenos Aires. [Influence of nearby oceans on seasonal average rainfall and temperature in Argentina. Bachelor thesis. Department of Atmosphere and Ocean Sciences, Faculty of Exact and Natural Sciences, University of Buenos Aires].

Oliveri, P. and González, M. H. (2018). La influencia de la temperatura de superficie del mar de los océanos cercanos sobre la temperatura media de invierno en Argentina. Preprint CONGREMET XIII, Buenos Aires. [Influence of the sea surface temperature of nearby oceans on the average winter temperature in Argentina].

Paegle, J. and Mo, K. C. (2002). Linkages between Summer Rainfall Variability over South America and Sea Surface Temperature Anomalies. *J. Climate* 15: 1389 – 1407.

Perez Pardo, O. (2002). *Manual on desertification*. Ministry of Environment and Sustainable Development, of the Nation (https://www.iapg.org.ar/

sectores/olimpiadas/certamenes/listados/2011/Desertificacion/Manuals obreDesertificacionenlaRA.pdf).

Reboita M. S., Ambrizzi T. and da Rocha R. P. (2009). Relationship between the southern annular mode and Southern Hemisphere atmospheric systems. *Rev Bras Meteorol* 24:48–55.

Rolla, A., Nuñez, M., Guevara, H., Meira, S., Rodriguez, G., Ortiz de Zárate and M. I. (2017). Climate Impacts on Crop Yields in Central Argentina. Adaptation strategies. *Agricultural Systems.* 160 (2018). https://doi.org/10.1016/j.agsy.2017.08.007.

Salio, P., Nicolini, M. and Saulo, A. C. (2002). Chaco low-level jet events characterization during the austral summer season. *Journal of geophysical research*, vol. 107, no. d24, 4816. doi:10.1029/2001jd001315.

Satyamurty, P., Mattos, L. F., Nobre, C. A. and Silva Dias, P. L. (1998). Tropics - South America. In: *Meteorology of the Southern Hemisphere,* Ed. Kauly, D. J. and Vincent, D. G., Meteorological Monograph. American Meteorological Society, Boston, 119-139.

Saurral, R., Inés A. Camilloni and R. Barros: 2016. Low-frequency variability and trends in centennial precipitation stations in southern South America, *Int. J. Climatol.,* Wiley Online Library, DOI: 10.1002/joc.4810.

Silvestri, G. and Vera, C. (2003). Antarctic Oscillation signal on precipitation anomalies over southeastern South America. *Geophys Res Letters* 30, 21: 21-15.

Stoll Villarreal, L., Castañeda, M. E. (2018). Pronóstico estadístico de agua en el suelo en Tres Arroyos, Provincia de Buenos Aires. Preprint CONGREMET XIII, Buenos Aires. [Statistical forecast of soil water in Tres Arroyos, Buenos Aires province. CONGREMET XIII Preprint, Buenos Aires].

Sudar, R. A., Saxton, K. E and Spomer, R. G. (1981). A predictive model of water stress in corn and soybeans. *Trans. of Am. Soc. Agric. Engr.* 24(1): 97-102.

Sun, X., Cook, K. H. and Vizy, E. K. (2017). The South Atlantic subtropical high: climatology and interannual variability. *J. Clim.* 30, 3279–3296. doi: 10.1175/JCLI-D-16-0705.1.

Thompson, D. W. and Wallace, J. M. (2000). Annular modes in the extratropical circulation. Part I: month-to-month variability. *J. Climate*, 13, 1000-1016.

UNCCF, 3CN. (2015). Third Communication of the Argentine Republic to the United Nations Framework Convention on Climate Change. (https://unfccc.int/sites/default/files/resource/Argnc3.pdf).

Vera, C., Silvestri, G., Barros, V. and Carril, A. (2004). Differences in El Niño response in the Southern Hemisphere. *J. Climate* 17, 9: 1741-1753.

Vera, C., Baez, J., Douglas, M. Emmanuel, C. B., Marengo, J., Meitin, J., Nicolini, M., Nogues-Paegle, J., Paegle, J., Penalba, O., Salio, P., Saulo, C., Silva Dias, M. A., Silva Dias, P. and Zipser, E. (2006). *The South American Low-Level Jet Experiment.* https://doi.org/10.1175/BAMS-87-1-63.

Wilks, D. (1995). *Satistical Methods in the Atmospheric Sciences, An Introduction.* Academic Pres.

INDEX

A

access, 4, 27, 31, 40, 43, 46, 51, 74
agricultural land use, vii, viii, ix, 25, 26, 32, 34, 43, 52, 53, 62
agricultural policy, 26, 43, 54
agricultural producers, 33, 35, 48, 50
agricultural sector, ix, x, 26, 33, 40, 45, 52, 54, 83
agriculture, x, 3, 26, 29, 30, 33, 34, 37, 40, 43, 44, 45, 47, 48, 52, 53, 54, 56, 58, 59, 60, 62, 71, 77, 80, 84, 88, 102
Argentina, 83, 84, 89, 95, 101, 102, 103, 104, 105, 106
assessment, 3, 10, 14, 22, 75, 79
assimilation, 103
atmosphere, 18, 19, 76, 90, 105
authorities, 38, 46, 48, 49, 51

B

bankruptcy, 39
barriers, 44, 54
basic needs, 3

benefits, 14, 19, 31
biodiversity, vii, ix, 4, 10, 11, 14, 19, 52, 60, 62, 66, 74
biological, 2, 6, 11, 62, 78, 79
Brazil, v, ix, 44, 59, 60, 63, 64, 66, 67, 69, 73, 74, 75, 76, 77, 80
breadth, 15
burn, 69
businesses, 42, 43
buyer, 72

C

capital accumulation, 31
carbon, vii, ix, 5, 18, 19, 60, 62, 67, 73
carbon dioxide, 5, 18
case study, ix, 26
cattle, ix, 60, 62, 63, 66, 67, 70, 71, 73
central executive, 51
century, 3, 5, 7, 14, 18, 42, 61
challenges, 5, 9, 15, 18, 43, 61, 62
change detection, 2, 17, 19
circulation, 89, 102, 104, 107
cities, 4, 5, 6, 12, 16, 20
citizens, 49

citizenship, 44
civil society, 40, 42, 43, 47, 48, 49
classification, 67, 101
climate, viii, 2, 6, 9, 15, 18, 19, 22, 61, 62,
        63, 68, 76, 79, 80, 84, 85, 89, 101, 102,
        103, 104, 105, 106, 107
climate change, 6, 15, 18, 19, 62, 68
climate XE "climate" forcing, 84
coatings, 4
collisions, 13
colonization, 63, 68, 69, 70
commercial, 27, 51, 66
commercial crop, 27
communication, 11, 14, 31
communities, 27, 47, 51, 52
community, 8, 12, 14, 39, 49
competitiveness, 27
complex interactions, 14
computer systems, 4, 10
configuration, 103
corporate sector, 26
corporatization of agriculture, 26
correlation, vii, x, 84, 89, 91, 92, 94
correlation coefficient, vii, x, 84, 91, 92
corruption, 28, 46, 55
crop, 49, 61, 63, 85, 88, 102
crop production, 61, 85, 88
cross-validation, vii, x, 84, 92, 93, 96, 98,
        99
cultural heritage, 14, 63

D

data processing, 4
deficit, 3, 85, 99, 101
deforestation, viii, x, 1, 4, 8, 19, 60, 62, 63,
        68, 69, 76, 77, 78, 79
degradation, viii, 2, 4, 6, 8, 19, 20, 45, 54,
        75, 85, 104
degradation process, viii, 2
dehumidification, 9

depth, 43, 65
deregulation, 51
desertification, vi, viii, 2, 4, 20, 83, 85, 105
destruction, viii, 2, 4, 5, 6, 16, 20
developing countries, 2, 5, 16
development, viii, ix, 2, 3, 4, 5, 6, 12, 15,
        16, 20, 22, 23, 26, 43, 44, 49, 52, 55, 62,
        63, 69, 70, 79, 103, 105
digitization, 4
displacement, 52, 95
distribution, x, 9, 30, 33, 34, 36, 44, 60, 71,
        90, 91, 93
drainage, 68, 102
drying, 16, 17, 90
dynamic, 3, 8, 10, 23, 62

E

early warning, 22
ecological processes, 70
ecology, 4, 12
economic cycle, 69
ecosystem services, vii, ix, 14, 60, 62, 74
ecosystems, 3, 5, 6, 20, 52, 61, 75
electrical conductivity, 68
employees, 27
employment, 61
employment opportunities, 61
entrepreneurs, 48
environment, 6, 14, 15, 18, 27, 31, 37, 46,
        61
environmental change, vii, 1
environmental crisis, 16
environmental degradation, 2, 6
environmental factors, 4
environmental impact, viii, 2
environmental issues, 2, 7, 9
environmental protection, 7, 18
erosion, viii, 2, 4, 9, 20, 61, 85
ethnic groups, 69
Europe, 42, 43, 55, 57, 72

European Parliament, 42, 43, 55, 57, 58
evapotranspiration, 9, 61, 68, 90, 102
evidence, ix, 26, 33, 45, 46
expansion, ix, 6, 60, 61, 62, 68, 71, 74, 79
Expansion, 71, 79
exploitation, 20
extraction, 10, 30

F

family farming, ix, 26, 54
farmers, 29, 30, 39, 40, 41, 42, 43, 44, 45, 47, 49, 54, 65, 69, 71, 72
farmland, 30, 33, 34, 35, 42, 44
farms, 3, 26, 27, 28, 29, 33, 34, 35, 39, 40, 42, 43, 44, 46, 54, 65, 66, 71
financial, 27, 31, 37, 39, 42, 44, 46, 53
financial institutions, 39, 42, 44
financial resources, 31, 46
flood, viii, 1
food, 2, 30, 34, 44, 52, 54, 61, 62
food production, 62
food security, 2, 44, 54
force, 28, 37, 38, 41, 44, 53
forecasting, 4, 93
forecasting model, 4
foreign investment, 41, 45
funds, 44, 50

G

Geographic Information System (GIS), v, ix, 1, 7, 9, 11, 22, 60, 63, 65
geoinformatics, 3, 21
Geoinformatics, 21
global, 2, 6, 7, 8, 18, 19, 22, 30, 35, 36, 37, 41, 47, 54, 55, 57, 61, 80, 84, 91
Global, 19, 30, 35, 54, 55, 57, 80
goods and services, 61, 62
governance, 46, 53, 65
grassroots, 30

greenhouse, 5, 18, 19, 79
greenhouse gas, 5, 18, 19
greenhouse gas emissions, 5, 18, 19
greenhouse gases, 18
Gross Domestic Product, 72
growth, viii, 2, 4, 5, 9, 12, 15, 16, 18, 20, 22, 23, 42, 61, 66, 67, 72, 73, 85, 103
growth rate, 5

H

habitat, 4, 5, 10, 12, 14, 62
habitat quality, 14
habitats, 5, 12, 14
harassment, 39
harvesting, 40, 49
hazards, 16, 18, 20
hemisphere, 105
history, 6, 12, 17
human, viii, 1, 3, 6, 10, 12, 14, 15, 18, 20, 30, 31, 53, 61, 62, 63
human health, 15
human right, 30, 31, 53
humidity, 91, 95, 101

I

ideal, 3
identification, 3
image, 4, 10, 16, 45, 65
imagery, 11, 16, 20, 21
images, vii, viii, 2, 17, 67
individuals, 32, 38
industrial revolution, 19
industries, 8
infrastructure, x, 46, 53, 60, 62, 63, 69
institutional change, 37
institutions, 42, 48
International Atomic Energy Agency, 75
International Monetary Fund, 54
intervention, 6, 18, 85

intimidation, 39
invertebrates, 12
investment, 44, 45
investments, ix, 26, 44, 45, 46, 52, 54
investors, 33, 36, 37, 44, 45, 46, 48, 52
irrigation, 17, 72, 102

J

judicial power, 37

K

Kazakhstan, 17

L

lakes, 16, 17
land acquisition, 36, 52, 71
land concentration, viii, ix, 26, 27, 29, 30, 31, 32, 33, 35, 36, 37, 41, 42, 43, 45, 46, 47, 48, 51, 52, 53, 54
land grabbing, viii, ix, 26, 27, 29, 30, 31, 32, 33, 37, 41, 42, 43, 45, 46, 47, 48, 49, 50, 51, 52, 53
land managenment, 60
land raiding, 26, 32, 33, 37, 39, 45, 48, 52, 53
land use, vii, viii, ix, 1, 3, 4, 8, 9, 10, 11, 14, 18, 20, 21, 22, 25, 26, 31, 33, 34, 38, 39, 41, 42, 43, 51, 52, 53, 54, 60, 61, 62, 63, 66, 67, 68, 70, 71, 72, 73, 75
Land Use /Land Cover (LULC), 2, 14
landscape, 4, 5, 6, 7, 12, 14, 21, 22, 23, 62, 63, 65, 69, 70, 71, 75, 77
land-use, viii, 2, 4, 5, 8, 9, 10, 17, 19, 20, 21
large-scale farming, 26
laws, 18, 39, 50, 56
lead, viii, 2, 20, 29, 31, 54, 61
legislation, 28, 29, 37, 39, 47, 50, 51, 66

linear model, 101
local government, 51
location information, 3
lower prices, 72

M

management, vii, ix, x, 1, 2, 5, 8, 9, 14, 15, 16, 20, 31, 43, 46, 47, 50, 51, 60, 61, 63, 65, 66, 67, 68, 69, 73, 74, 83
manpower, 71
mapping, vii, viii, 2, 4
marginalization, 6, 44
market failure, 46
mass, 37
measurements, x, 14, 60, 89
methodology, 74, 96, 98, 101
migration, 12
migration routes, 12
military, 51
models, vii, x, 4, 5, 7, 9, 17, 19, 65, 84, 92, 93, 96, 98, 99, 100, 101
modernization, 62
modifications, 37, 61, 90
moisture, 4, 103
moratorium, 27, 28, 29
multi-sensory data, 11
multi-spectral, 4

N

National Academy of Sciences, 25, 78, 79
natural resources, vii, 1, 3, 9, 16, 20, 47
natural tourism, 8
negative effects, ix, 6, 8, 26
neural network, 101
neural networks, 101
nutrient, 61, 62, 67, 74

O

oceans, 105
oil, 6, 8, 71
opportunities, 29, 33, 45, 46, 52, 53
organic matter, 62, 85
ownership, 30, 31, 32, 41, 46, 49

P

Pacific, 13, 95, 101
Parliament, 28, 29, 43
passageway, 11
Passageway, 11
pasture, 3, 66, 67, 69, 70, 71
pastures, ix, 60, 66, 67, 71, 73
percolation, 90
phonological, 4
physical health, 6
physical structure, 68
policy, vii, ix, 26, 33, 43, 48, 54, 60, 65, 69
policy makers, vii, ix, 60
political instability, 37
political power, 31
pollution, viii, 1, 6, 15, 18, 20
population, viii, 2, 6, 11, 31, 47, 61, 69
population growth, viii, 2
precipitation, 62, 68, 84, 85, 86, 87, 89, 90, 95, 103, 105, 106
predictability, 102
prediction models, 4, 5, 8, 9
preparation, iv
preservation, 14, 43, 49, 75
principles, ix, 23, 26, 52, 104
probability, 93, 101
production targets, 74
property rights, 39, 50
protection, 9, 14, 19, 46, 48, 51
public policy, 47, 66

Q

qualitative, 3, 11
quality of life, 2
quantitative, 3, 11
quotas, 30

R

rain forest, ix, 60, 62, 63, 75
rainfall, 61, 64, 68, 89, 102, 104, 105
redistribution, 37, 51
reflexes, 62
reform, 29, 46, 51
regional economies, 85
registries, 38
regression, 90, 92, 96, 101
remote sensing, v, x, 1, 2, 3, 4, 9, 11, 16, 19, 20, 21, 22, 23, 60, 76, 77
requirements, 28, 51, 102
researchers, 20, 42
reserves, 10, 65, 69, 71
resources, vii, viii, 1, 2, 4, 5, 9, 16, 20, 27, 32, 39, 47, 50, 61, 70
rights, 28, 31, 32, 38, 39, 40, 41, 45, 46, 47, 49, 51, 52, 54
risks, 9, 47
routes, 12, 13
runoff modeling, 4
rural areas, ix, 26, 29, 33, 52, 53, 54
rural development, ix, 26, 49, 52

S

salinity, 3
satellite, 3, 16, 17, 20, 22, 65, 67, 75
saturation, x, 60, 65
Saudi Arabia, 6, 36
savannah, 63
savings, 19

schema, 69
seasonal forecast, x, 83, 84
security, 22, 39, 46, 53
sediment, viii, 1, 9
self-sufficiency, 52
sensing, 3, 9, 11, 19, 20, 21
sensitivity, 5, 14, 93, 94
sensory data, 11
settlement policy, 70
settlements, 6, 12, 68, 69, 70
simulation, 9
social consequences, ix, 26
socioeconomic, viii, 2
soil functions, vii, ix, 60, 62, 63, 65, 67, 75
soil type, 65
soil water, vii, x, 83, 84, 85, 88, 89, 91, 99, 101, 106
South America, 95, 102, 103, 104, 105, 106, 107
species, 3, 12, 14, 88
stakeholders, x, 60, 65, 71, 74
state, ix, 8, 26, 27, 29, 38, 41, 43, 46, 48, 50, 51, 53, 60, 72
state-owned enterprises, 51
states, 44, 49
statistics, 12, 33, 34, 35, 58, 75, 77
storage, vii, x, 60, 84, 90
sustainable development, 5, 17, 20, 52
symbiosis, 69

T

takeover, 31, 32, 40
target, x, 60, 70
tax deduction, 70
technical support, 69
techniques, 4, 11, 17, 20
technologies, 3, 9, 46
temperature, 18, 61, 63, 84, 85, 87, 89, 91, 95, 101, 102, 105
tenure, 53, 65, 71

terrestrial, 4, 5, 6
territorial, 47, 69
territory, 30, 69, 84
texture, 65, 67
threats, 14, 43, 44, 54, 62
time frame, 66
time series, viii, 2, 11, 14, 66, 67, 68
tones, 61
topography, viii, 2, 63
tourism, 8, 16
trade, 61, 63, 72
trading partners, 72
training, 92
transactions, ix, 26, 27, 29, 33, 36, 44, 52
transformation, 5, 6, 11, 37
transparency, 36
transport, 61, 70, 94, 95

U

Ukraine, vii, viii, ix, 25, 26, 27, 28, 29, 31, 32, 33, 34, 35, 36, 37, 39, 41, 44, 45, 46, 47, 48, 49, 50, 51, 52, 53, 56, 57, 58
United Nations, 19, 31, 47, 77, 84, 102, 107
United Nations Framework Convention on Climate Change, 19, 84, 107
United States, 12, 13, 14, 36, 78, 79
urban, viii, 1, 5, 7, 9, 22, 23, 67, 71
urban areas, 9
Urmia Lake, 5, 15
Uzbekistan, 17

V

validation, 92, 98
valuation, 51
variables, 89, 90, 91, 94
vast areas, 5, 16
vegetation, ix, 4, 6, 10, 60, 61, 64, 70, 71, 85
violence, 32

W

water, vii, ix, x, 2, 3, 6, 9, 15, 16, 17, 22, 60, 61, 62, 66, 67, 68, 73, 83, 84, 85, 88, 89, 90, 91, 99, 101, 102, 106
water purification, 62

water quality, 68
water resources, x, 2, 6, 9, 15, 22, 84
well-being, 61
wetlands, 16
wildlife, 5, 11, 12, 14, 15
Word, vi, 46, 55, 58, 85, 88, 109

Related Nova Publications

PEARS: CULTIVARS, PRODUCTION AND HARVESTING

EDITOR: Alberto Ramos Luz

SERIES: Agriculture Issues and Policies

BOOK DESCRIPTION: This book brings up-to-date information on different topics about the pear tree cultivation. Basic content and more specific and in-depth content are presented such as a series of research results and experiences on behavior and management tools to grown pear trees in subtropical climate, warmer conditions of the traditionally cultivated areas, subject of world-wide interest in the face of the climatic changes that are occurring over the years.

HARDCOVER ISBN: 978-1-53616-036-9
RETAIL PRICE: $160

GLOBAL AGRICULTURAL EXTENSION PRACTICES: COUNTRY BY COUNTRY APPROACHES

EDITORS: Dixon Olutade Torimiro and Chris Orobosa Igodan

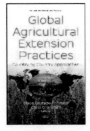

SERIES: Agriculture Issues and Policies

BOOK DESCRIPTION: Agricultural Extension practices or advisory services as per usage in some countries provide a close examination of country-by-country approach. In the book, contributions are drawn from thirteen countries in four regions of the world.

HARDCOVER ISBN: 978-1-53616-012-3
RETAIL PRICE: $230

To see a complete list of Nova publications, please visit our website at www.novapublishers.com

Related Nova Publications

COTTON: HISTORY, PROPERTIES AND USES

EDITOR: Jules Dagenais

SERIES: Agriculture Issues and Policies

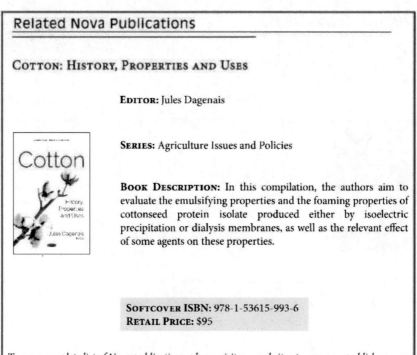

BOOK DESCRIPTION: In this compilation, the authors aim to evaluate the emulsifying properties and the foaming properties of cottonseed protein isolate produced either by isoelectric precipitation or dialysis membranes, as well as the relevant effect of some agents on these properties.

SOFTCOVER ISBN: 978-1-53615-993-6
RETAIL PRICE: $95

To see a complete list of Nova publications, please visit our website at www.novapublishers.com